Intelligent Systems Reference Library

Volume 111

Series editors

Janusz Kacprzyk, Polish Academy of Sciences, Warsaw, Poland
e-mail: kacprzyk@ibspan.waw.pl

Lakhmi C. Jain, University of Canberra, Canberra, Australia;
Bournemouth University, UK;
KES International, UK
e-mails: jainlc2002@yahoo.co.uk; Lakhmi.Jain@canberra.edu.au

About this Series

The aim of this series is to publish a Reference Library, including novel advances and developments in all aspects of Intelligent Systems in an easily accessible and well structured form. The series includes reference works, handbooks, compendia, textbooks, well-structured monographs, dictionaries, and encyclopedias. It contains well integrated knowledge and current information in the field of Intelligent Systems. The series covers the theory, applications, and design methods of Intelligent Systems. Virtually all disciplines such as engineering, computer science, avionics, business, e-commerce, environment, healthcare, physics and life science are included.

More information about this series at http://www.springer.com/series/8578

Eman El-Sheikh · Alfred Zimmermann
Lakhmi C. Jain
Editors

Emerging Trends in the Evolution of Service-Oriented and Enterprise Architectures

Editors
Eman El-Sheikh
Center for Cybersecurity
University of West Florida
Pensacola, FL
USA

Alfred Zimmermann
Hochschule Reutlingen
Reutlingen
Germany

Lakhmi C. Jain
KES International
Leeds
UK

and

University of Canberra
Canberra, ACT
Australia

and

Bournemouth University
Poole
UK

ISSN 1868-4394   ISSN 1868-4408   (electronic)
Intelligent Systems Reference Library
ISBN 978-3-319-82130-6   ISBN 978-3-319-40564-3   (eBook)
DOI 10.1007/978-3-319-40564-3

Printed on acid-free paper

This Springer imprint is published by Springer Nature
The registered company is Springer International Publishing AG Switzerland

Foreword

Almost 25 years ago Henderson and Venkatraman were writing *"Even though the information technology has evolved from its traditional orientation and administrative support toward a more strategic role within an organization, there is still a glaring lack of fundamental frameworks within which to understand the potential of IT for tomorrow's organizations"* (IBM Systems Journal 32(1), 1993). Targeting to master this challenging issue, they developed a model, strategic alignment model, and derived four perspectives of alignment with specific impacts for guiding management practices in order to *"leverage IT for transforming organizations."*

More recent technologies and paradigms (such as cloud computing, big data, IoT) show that not the ownership of IT resources but their strategic management is the foundation for sustainable competitive advantage, as it was earlier defended by Mata, Fuerst, and Barney (MIS Quaterly, 1995).

A decade ago, Ross, Weill, and Robertson in their book "Enterprise Architecture as Strategy: Creating a Foundation for Business Execution" (2006) illustrated through numerous companies worldwide, how constructing the right enterprise architecture enhances profitability and time to market, improves strategy execution, and even lowers IT costs.

Enterprise architecture (EA) aimed (i) to understand the interactions and all kinds of articulations between business and IT, (ii) to define how to align business components and IT components, as well as business strategy and IT strategy, and more particularly (iii) to develop and support a common understanding and sharing of those purposes of interest. EA is also used to map the enterprise goals and strategy to the enterprise's resources (actors, assets, and IT supports) and to manage the evolution of this mapping.

Services are the governing principle for EA. Nearly all newly created EAs are service-oriented. Service-oriented enterprise architecture (SoEA) easily integrates widespread technological approaches such as SOA or emerging ones as cloud computing because they also use service as structuring and governing paradigm. The scope of SoEA is much broader than the scope of the SOA and also includes services not accessible through software such as business services and infrastructure services. Services of different purposes and granularities may be interconnected in service (value) nets to provide higher-level services.

Today, foundations of social computing influence EA in new ways. The senior management defines organizational structures no longer alone, but weak ties that are initiated by individuals superimpose the organization. Innovation is no longer a process guided by an elite, but can be initiated by every member of an organization. Decisions are no longer only made by experts, but are also results of collaborative processes. Big data technologies allow to process data with higher velocity, variety, and volume and to create new information flows and data services within EAs.

EA is positioned as a coordination and steering mechanism and as an instrument to support the strategic direction of digital enterprises, which new frontiers require permeability and which new structures require elasticity. The service paradigm and the underlying mechanisms offer an accelerator for nurturing the elasticity of EAs and that of the enterprises themselves, to allow them to survive in evolving business ecosystems. In this context, service ecosystems offer a new land of application for the Nash equilibrium.

The new challenges for the "design by reuse" of modern IT solutions (recommended to be built in shorter cycles), in accordance with SOA and EA frameworks, impose in turn new challenges to the "design for reuse." The latter should (i) handle the potential components (services), in terms of abilities to satisfy functional business requirements in manyfold contexts and also (ii) deal with new capabilities for mastering nonfunctional requirements, such as flexibility, maintainability, and trustworthiness, which may themselves be variable in different contexts.

The twelve chapters of this book all together present challenging issues and hot topics related to the emerging trends in the evolution of service-oriented and enterprise architectures, as the evolution of EAs and systems, the flexibility, the maintainability, the security of the underlying software solutions and infrastructures, the digital transformation, the capability management, the forecasting of service demands, the conciliation of resilient and stable parts of EA, which are essential for the integrity of transactions and reliability of systems, with a fast-speed-architecture offering channels that are pivotal for the customer experience. As advocated by one of the contributors, *"Digital Transformation sets a new challenge for the enterprise architect: she has now not just to align the IT with the demands from the business but to enable and even invent new business opportunities. So the architecture capability of an organization gets an active part in shaping the business"*.

Selmin Nurcan
University Paris 1 Panthéon-Sorbonne

Preface

This research oriented book presents emerging trends in the evolution of Service-Oriented and Enterprise architectures. New architectures and methods of both business and IT are integrating services to support mobility systems, Internet of Things, Ubiquitous Computing, collaborative and adaptive business processes, Big Data, and Cloud ecosystems. They inspire current and future digital strategies and create new opportunities for the digital transformation of next digital products and services. Service-Oriented Architectures (SOA) and Enterprise Architectures (EA) have emerged as useful frameworks for developing interoperable, large-scale systems, typically implementing various standards, like Web Services, REST, and Microservices. Managing the adaptation and evolution of such systems presents a great challenge. Service-Oriented Architectures enable flexibility through loose coupling, both between the services themselves and between the IT organizations that manage them. Enterprises evolve continuously by transforming and extending their services, processes and information systems. Enterprise Architectures provide a holistic blueprint to help define the structure and operation of an organization with the goal of determining how an organization can most effectively achieve its objectives. This book presents several novel approaches to address the challenges of the service-oriented evolution of digital enterprise and software architectures.

The book is directed to the researchers, postgraduate, graduate and undergraduate students, professors and practitioners who are interested in the service-oriented evolution of digital enterprise and software architectures.

We are grateful to the contributors and reviewers for their very valuable expertise and contributions without which this book would not have existed. We wish to show our appreciation to Springer-Verlag for their support right from the concept development to the final typesetting phase of this book.

The unconditional support provided by our universities is acknowledged.

USA   Eman El-Sheikh
Germany   Alfred Zimmermann
Australia   Lakhmi C. Jain

Contents

About the Editors

Eman El-Sheikh is professor of computer science and director of the Center for Cybersecurity at the University of West Florida. She received her Ph.D. and M.Sc. in computer science from Michigan State University and her B.Sc. in computer science from the American University of Cairo. Her research interests include artificial intelligence, machine learning, intelligent systems, cybersecurity and software maintenance and evolution. Dr. El-Sheikh has applied her research to various problem domains, including education, health care, finance, software maintenance, and robotics. She has authored over 60 publications and given over 50 conference presentations in her research areas and enjoys engaging and mentoring undergraduate and graduate students in research activities.

Alfred Zimmermann is a professor of computer science at Reutlingen University, director of the Graduate Cooperative Research School for Services Computing, and the research director of the Herman Hollerith Center, Boeblingen, Germany. His research is focused on digital transformation and digital enterprise architecture in close relationship with services and cloud computing. He graduated in medical informatics at the University of Heidelberg and obtained his Ph.D. in informatics from the University of Stuttgart, Germany. He keeps the academic relations of his home university to the GI——the German Computer Science Society, the ACM——the US Association for Computing Machinery, and the IEEE, where he is a part of specific research groups, programs, and initiatives like software architecture, enterprise architecture and management, services computing, and cloud computing. Additionally, he is a visiting professor and honorary professor at international universities: La Plata University—Buenos Aires, the Marmara University, and the Yeditepe University of Istanbul.

Lakhmi C. Jain is a visiting professor at Bournemouth University, UK, and adjunct professor at University of Canberra, Australia.

Dr. Jain founded the KES International for providing a professional community the opportunities for publications, knowledge exchange, cooperation, and teaming. Involving around 5000 researchers drawn from universities and companies worldwide, KES facilitates international cooperation and generates synergy in teaching and research. KES regularly provides networking opportunities for professional community through one of the largest conferences of its kind in the area of KES.

His interests focus on the artificial intelligence paradigms and their applications in complex systems, security, e-education, e-health care, unmanned air vehicles, and intelligent systems.

www.kesinternational.org

Chapter 1
Evolution of Service-Oriented and Enterprise Architectures: An Introduction

Eman El-Sheikh, Alfred Zimmermann and Lakhmi C. Jain

Abstract This chapter presents an introduction to emerging trends in the evolution of service-oriented and enterprise architectures. The primary aim of this book is to highlight some of the most recent research results in the field. Brief descriptions of the chapters included in the book are provided.

1.1 Introduction

Services Oriented Architectures (SOA) and Enterprise Architectures (EA) have emerged as useful frameworks for developing interoperable, large-scale systems, typically implemented using the Web Services (WS) standards [1]. SOA typically refers to large systems-of-systems in which composite applications are created by orchestrating loosely coupled service components that run on different nodes and communicate via message passing [2]. Often an infrastructure layer, sometimes called an Enterprise Service Bus (ESB), mediates the communication, providing features such as routing, security, and data transformation. Such systems present several software engineering challenges because they need to orchestrate diverse services having different owners, and have complex reliability requirements.

While developing SOA applications presents many software engineering challenges, managing the evolution of such systems presents even greater challenges [3, 4]. SOA

E. El-Sheikh (✉)
Center for Cybersecurity, University of West Florida, Pensacola, FL, USA
e-mail: eelsheikh@uwf.edu

A. Zimmermann
Faculty of Informatics, Reutlingen University, Reutlingen, Germany
e-mail: alfred.zimmermann@reutlingen-university.de

L.C. Jain
University of Canberra, Canberra, Australia
e-mail: Lakhmi.Jain@canberra.edu.au; jainlc2002@yahoo.co.uk

L.C. Jain
Bournemouth University, Poole, UK

© Springer International Publishing Switzerland 2016
E. El-Sheikh et al. (eds.), *Emerging Trends in the Evolution of Service-Oriented and Enterprise Architectures*, Intelligent Systems Reference Library 111,
DOI 10.1007/978-3-319-40564-3_1

1

provides flexibility through loose coupling, both between the services themselves and between the IT organizations that manage them. However, this flexibility may create significant problems when the composite application needs to evolve. Several researchers have pointed out challenges in the evolution of SOA systems [3, 4, 5].

Information, data and knowledge are fundamental concepts of our everyday activities. Social networks, smart portable devices, and intelligent cars, represent only a few instances of a pervasive, information-driven vision for the next wave of the digital economy and the digital transformation. Digitization has a major impact for the current development of modern societies and questions fundamental structures in society, economy, and technology. Smart connected products and services expand physical components from their traditional core by adding information and connectivity services using the Internet. Digitized products and services with service-oriented architectures amplify the basic value and capabilities and offer exponentially expanding opportunities. Both business and technology are impacted from the digital transformation by complex relationships between architectural elements, which directly affect the evolution of adaptable service-oriented architectures for digital products and services and their related digital governance.

Enterprise Architectures provide a conceptual blueprint to help define the structure and operation of an organization with the goal of determining how an organization with their digitized services can most effectively achieve its current and future objectives. Several approaches and methods have been proposed to address the challenges of EA evolution and manage digital transformations. Enterprise architecture comprises a holistic, consistent and coherent set of principles, methods, guidelines, and models, which are used to support the top level design and implementation of organizational structures, business products and services, business processes, information systems, and their infrastructure. Enterprise architecture is positioned as a coordination and steering mechanism and as an instrument to support the digital enterprise's strategic direction.

This book presents emerging trends in the evolution of service-oriented and enterprise architectures. Chapter 2 describes how SOA design principles can facilitate SOA evolvability and examines several approaches and emerging trends to support and enhance SOA evolution. Chapter 3 presents a toolset that developers can use to design resource-oriented web services in a service-oriented architecture systematically in a quality-oriented manner. Chapter 4 describes a knowledge elicitation and modeling approach to identify trust and security concerns as SOA systems evolve along with two examples of knowledge modeling in support of SOA system evolution. Chapter 5 describes the fractal nature of SOA designs for sustainment management tools as these tools evolve into even more dynamic, federated systems and summarizes insights gained from more than twenty years of software development, maintenance, and evolution of a major pavement engineering tool named PAVER™. Chapter 6 investigates mechanisms for analyzing enterprise architectures to provide decision support for architectural evolution and adaptation and presents a novel approach that leverages a new extended digital enterprise architecture model that is well suited for adaptive models and transformation mechanisms.

Chapter 7 describes a lightweight enterprise architecture framework that provides enterprise architects with an agile development approach for digital transformations. Chapter 8 presents a two-speed enterprise architecture that enables more established companies to manage digital transformation. Chapter 9 describes a novel approach for designing capabilities to tackle the challenges of rapidly changing enterprise environments by modeling the application context. Chapter 10 reviews the body of capability-driven management literature and provides an overview of capability research investigations over the last 15 years. Chapter 11 highlights the increased capabilities of enterprise architecture analytics and decision support through the use of a data-driven approach and provides insights into current research work in this area. Chapter 12 describes a capability management guide that provides a flexible "engineering" approach for identifying, structuring, and maintaining enterprise capabilities.

References

1. Josuttis, N.M.: SOA in Practice: The Art of Distributed System Design. O'Reilly. ISBN: 0-596-52955-4 (2007)
2. Lewis, G., Morris, E., Simanta, S., Smith, D.: Service orientation and systems of systems. IEEE Softw. **28**(1), 58–63 (2011). doi:10.1109/MS.2011.15
3. Gold, N., Knight, C., Mohan, A., Munro, M.: Understanding service-oriented software. IEEE Softw. **21**, 71–77 (2004). doi:10.1109/ms.2004.1270766
4. Lewis, G.A., Smith, D.B.: Service-oriented architecture and its implications for software maintenance and evolution. Frontiers Softw. Maint. 2008 (FoSM), 1–10. doi:10.1109/fosm. 2008.4659243
5. Gold, N., Bennett, K.: Program comprehension for web services. International Conference on Program Comprehension (2004). doi:10.1109/wpc.2004.1311057

Chapter 2
Approaches to the Evolution of SOA Systems

Norman Wilde, Bilal Gonen, Eman El-Sheikh
and Alfred Zimmermann

Abstract The evolution of Services Oriented Architectures (SOA) presents many challenges due to their complex, dynamic and heterogeneous nature. We describe how SOA design principles can facilitate SOA evolvability and examine several approaches to support SOA evolution. SOA evolution approaches can be classified based on the level of granularity they address, namely, service code level, service interaction level and model level. We also discuss emerging trends, such as microservices and knowledge-based support, which can enhance the evolution of future SOA systems.

2.1 Introduction

Early in the history of modern computing it became evident that most of the software developer's work actually takes place after an application's initial delivery. This work came to be known as "software maintenance". Despite its economic importance, in the literature it was usually relegated to a supposedly uninteresting box at the bottom end of the waterfall software development life cycle.

With time, the term "maintenance" became unpopular because it was found that most of the work had little to do with repair, and much to do with the evolution of user needs and of computing environments [1]. As each new need or environment emerges the application must either adapt, be rewritten, or die.

N. Wilde
Department of Computer Science, University of West Florida, Pensacola, FL, USA
e-mail: nwilde@uwf.edu

A. Zimmermann
Faculty of Informatics, Reutlingen University, Reutlingen, Germany
e-mail: alfred.zimmermann@reutlingen-university.de

B. Gonen
School of Information Technology, University of Cincinnati, Cincinnati, OH, USA

E. El-Sheikh (✉)
Center for Cybersecurity, University of West Florida, Pensacola, FL, USA
e-mail: eelsheikh@uwf.edu

© Springer International Publishing Switzerland 2016
E. El-Sheikh et al. (eds.), *Emerging Trends in the Evolution of Service-Oriented and Enterprise Architectures*, Intelligent Systems Reference Library 111,
DOI 10.1007/978-3-319-40564-3_2

So today we speak more of "software evolution" than "software maintenance". This term is also somewhat problematic since "to evolve" is a passive verb, and thus gives the impression that evolution is something that just happens. In fact, keeping an application up to date requires very hard work, often under cruel deadline pressure, performed by very highly qualified professionals. In this chapter we use both terms since either, "evolution" or "maintenance", allows us to distinguish a greenfield software development situation in which design decisions can be taken freely, from the highly constrained context faced in making changes to an existing system.

The defining characteristic of software maintenance/evolution as opposed to development is that any proposed change needs to take into account a large base of existing software. This software has usually been molded by decisions taken, possibly years earlier, in circumstances very different from the current reality. Any change is thus highly constrained.

The emergence of Services Oriented Architecture (SOA) systems in the first decade of this century certainly did not eliminate the problems of software evolution, but it did change their nature. As a series of authors have pointed out, some aspects of SOA favor the job of the maintainer while others make it more difficult. New challenges are created both for practitioners and for researchers [2–6].

In this chapter we first briefly discuss perspectives on software evolution in general before going on to highlight some of the main approaches when these perspectives are applied to SOA systems. We cannot attempt to identify all of the diverse approaches to our subject, but we aim to contrast some of the main themes and explore advantages and disadvantages. The books and papers we mention are by no means an exhaustive list, but rather typify different ways of looking at the SOA evolution problem. We close with some discussion of emerging trends both in SOA architectures and in using knowledge-based methods to understand these architectures.

2.2 Perspectives on Software Evolution

One can identify two broad perspectives on software evolution that have dominated both theory and practice. On the one hand, software can be designed initially to make evolution easier. This is a perspective for the original software developer. It focuses on design approaches and implementation architectures that are hoped to facilitate future evolution. We might call this approach *design for evolvability*.

The second perspective looks for tools and methods to support ongoing maintenance of an existing system. This is a perspective for a maintenance software engineer. He must accept the system as it is, warts and all, and try to do the best possible job of keeping it up to date. We could call this perspective *support for evolution*.

2.2.1 Design for Evolvability

In design for evolvability, a key theme has been to find the "right" modularity. Any large application must be implemented as modules that connect together to provide the overall system functionality. The choice of modules, their interfaces, and the connection methods all strongly affect the ease with which changes can be made.

A key initial insight was the concept of "information hiding", generally credited to Parnas writing in 1972:

> We propose instead that one begins with a list of difficult design decisions or design decisions which are likely to change. Each module is then designed to hide such a decision from the others [7].

A designer thus should anticipate change, and hide expected changes within a single module. If the maintainer has to make one of the expected changes, then he or she can work within that single module. Design, coding and testing of the change will thus be far simpler than they would have been if many modules had been affected.

The difficulty for the designer, of course, is in identifying the likely changes so early in the system's lifetime. Security and performance considerations can also constrain what data and functionality need to be kept together, and thus lead to a decomposition that may seem less than optimal.

It is also not trivial to find a way of implementing the resulting modules without encountering insuperable barriers in the programming language or runtime environment. Many of the advances of programming languages have involved providing better mechanisms for modularity, with an explicit or implied goal of facilitating information hiding.

2.2.2 Support for Evolution

A great diversity of tools has been proposed to support the evolution of an existing software application. One might think of configuration management systems, editors that compare software versions, impact analysis tools, regression testing frameworks, and so on. But one important theme has been to provide the maintainer with support for *program comprehension*.

If the biggest difference between development and evolution is the presence of an existing code base, then the biggest practical difference to the software engineer is his need to *understand* that code base. The developer presumably understands the code he deals with because he, or his immediate colleagues, wrote it. The maintainer is often separated from the code's original authors by a distance of many miles or many years. He or she must reconstruct sufficient understanding of the application to be able to change it safely, without unexpected and possibly disastrous side effects.

Understanding legacy software is a complex task, both because of the scale of many existing software applications and because of the variety of relationships that may need to be understood. It has long been clear that software maintainers cannot attempt to understand large applications in their entirety or each maintenance task would take far too long. Instead they try to use a pragmatic as-needed strategy to understand only what is immediately relevant for the task at hand [8]. Even within that limitation, however, there is a bewildering variety of information which may be relevant: the structure of program text, dynamic structure and control flow at runtime, program functionality and programming plans at different levels of abstraction, domain knowledge which relates concepts in the real world to structures in the code, and so on [9].

2.3 Design for Evolvability Approaches to SOA

Services Oriented Architecture is generally regarded not as a specific architecture, but rather as a general architectural style for structuring software applications. Terminology varies but typically *composite applications* are constructed by orchestrating *services* running on different *nodes* and communicating via message passing. Often an infrastructure layer, sometimes called an *Enterprise Service Bus* (ESB), mediates service interactions providing functions such as message routing, reliable messaging and data transformations (Fig. 2.1).

Within the broad constraints of this style, there are many different ways of architecting any particular application. Many commentators on SOA have enunciated sets of design principles to guide this process (e.g. [10]), and many of the principles have evolvability as a goal. Some of the principles have become common practice while others are still somewhat aspirational.

One general principle is *loose coupling* of services, meaning generally that the designer tries to reduce dependencies between services as much as is practicable. Loose coupling may aid evolution because changes that would otherwise have required the intervention of a maintenance software engineer are instead handled

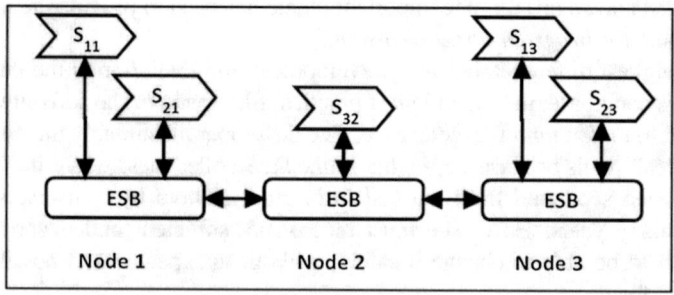

Fig. 2.1 Structure of a simple SOA composite application

automatically. For example at run time each service can publish its address in a registry, so that other services can find it automatically. The alternative of a hardwired address in code would require an edit, a build, and a new deployment. However as Josuttis points out, many of the more advanced strategies designed to provide loose coupling, such as asynchronous communication and error handling by compensation, have the side effect of increasing the complexity of code and thus may hinder maintainability [11].

Perhaps the most universally applied SOA design principle is that each service should implement a published interface or contract. This principle restates the module information hiding principle mentioned earlier; the service is a module that hides all the implementation details behind the interface. The Web Services Description Language (WSDL) [12] was developed to standardize interface descriptions across different hardware and software platforms. If the interface is unchanged, the service implementation can evolve with little or no impact on other services or on the application as a whole.

The WSDL standard greatly aids runtime linking of services, but it still leaves great flexibility in the design of services and their interfaces. These design decisions can have a great impact on evolvability. For example Borovskiy et al. [13] argue that generic services with flexible interfaces should often be preferred. However there is a tradeoff since a generic interface is less explicit about the data a service expects to receive, and thus provides less guidance to service consumers.

An alternative to formulating general principles for SOA design is to formulate design *patterns* to guide SOA design. This approach recognizes the difficulty of establishing principles that are universally applicable and instead defines patterns that have been found to be effective to achieve specific design goals in specific circumstances. These design goals often have to do with evolvability and specifically with managing versioning of a service as it changes (see [14], especially Chap. 7).

WSDL interface descriptions also aid evolution by facilitating interoperability of components from diverse owners. Interoperation has been described as having two or more independent systems operate in a coordinated and meaningful fashion such that processes are effectively merged or information is effectively shared [15]. If reusable, interoperable components are utilized, then both development and maintenance of the composite application should be easier because less special-purpose coding should be required.

However service reuse, as an organizational SOA strategy, seems to have been more difficult to achieve. Some companies have set out to build portfolios of reusable services with the idea that these will then be composed into new applications as the need arises. In some cases the concept seems to be that business process modeling tools will allow this composition to be done by business experts with little intervention from scarce software engineers. However the development of large-scale reusable software components has always been difficult [16] and the complexities of the assembly of distributed components are daunting. It is perhaps not surprising that Josuttis finds that assembly is best done "... by business and IT experts sitting together" and that service reuse is often less than expected [11]. In general the contribution of reuse to SOA evolution has probably been much less than originally anticipated.

2.4 Support for Evolution Approaches to SOA

As previously stated, the support for evolution perspective looks for tools and methods to aid in the ongoing maintenance of an existing system. The maintainer has to take the system as it is and solve problems effectively while working within organizational time constraints. A key theme in this perspective is helping the maintainer understand the existing system so that he or she can make changes safely.

As we look at SOA composite applications, perhaps one way to classify the different support for evolution approaches would be to look at the level of granularity they address. We could distinguish:

1. Service code level approaches that focus on understanding and manipulating the source code for a service.
2. Service interaction level approaches that focus on understanding how services work together in the application.
3. Model level approaches that focus on how the services relate to models of the application's domain.

2.4.1 Code Level Approaches to SOA Evolution

Historically, many organizations decided to begin their SOA efforts by exposing existing data or functionality as web services. Vendors soon moved to provide tools to automate this process so that it became easy to expose code written in a wide range of languages, from COBOL to C#, and hosted on platforms ranging from mainframes to Linux™.

The tools vary in their capabilities, but it is common to provide the ability to take existing code and create a service and its WSDL, or to take a WSDL and create shell code for a client or a service implementation. For example in the Java environment, one can take a Java class annotated with @WebService and use the *wsgen* tool to create the WSDL and classes that will handle the messaging. Going the other way, one can take a WSDL from an existing service and use the *wsimport* tool to create shell code for a client to access that service. Finally, if using a "WSDL-first" or "contract-first" development style, one may create the WSDL by hand and, once it is approved by all stakeholders, use it to generate shell code for the service implementation [17].

If a composite application was developed using source code based tools, then it is natural to continue maintaining it using these same tools. One advantage is that most of the vendor tools are available through an integrated development environment (IDE) that supports program comprehension with code search, code navigation and debugging facilities.

However a focus on code and code-generating tools for SOA evolution also has several pitfalls. If used incautiously, the tools can generate unwanted dependencies between a service and its client, thus tightening the coupling between them. To take just one example, a software engineer may accidentally include in an interface some of a service's internal data types. Then the client will necessarily have to use these same data types. Any change to the data structure on the server will force a change in the client. The more such code-generating tools are used over the life of a system, the more likely it is that this sort of design flaw will be introduced.

An important consideration may be the *depth* of the change. Papazoglou et al. [18] distinguish between shallow changes, that affect a single service and its immediate clients, and deep changes whose effects may cascade widely within a system. Perhaps a code focus may be acceptable in dealing with shallow changes, but deep changes require a more complex service life cycle to allow for more complete analysis and for more time for changes to propagate across the service landscape.

2.4.2 Service Interaction Level Approaches to SOA Evolution

Many of the tasks involved in SOA evolution do not require studying the code, but rather focus on understanding the interactions between services. For example a maintainer may be considering reconfiguring or replacing a service. Or he may need to locate where particular kinds of data are exchanged or where performance bottlenecks are developing. Much of the research on SOA maintenance and program comprehension has thus focused at the service interaction level of granularity.

It may be convenient to distinguish here between tools implementing dynamic and static approaches, since the practicalities of using tools will be different in each case. Dynamic tools get their input from actual execution of the system and use logs or message traces, often supplemented to meet the needs of the tool. Static approaches take as their input descriptions of the SOA system, such as requirements or design documentation, UML models, WSDL interface descriptions, etc. Each approach has its advantages and drawbacks.

One of the earliest dynamic approaches came from a group at IBM. De Pauw et al. [19] describe a visualization tool, Web Services Navigator, which helps users to understand SOA applications better. The tool collects data from event logs and processes it to generate visual abstractions, such as flow patterns, as well as views of transaction flows and data content. The paper describes how the tool has been used to understand overall application behavior in several different problem solving scenarios.

A narrower application of dynamic analysis attempts to recover and understand feature sequences, that is, the service interaction messages that occur when an end user makes use of a particular feature of the application. The problem is that there may be other concurrent users or routine system interactions that obscure the desired feature. Coffey et al. compute a relevance index for each observed message, giving greater weight to messages that are seen when a feature is known to be active. A Feature Sequence Viewer lets the maintainer set a threshold to view the sequence of the most relevant messages [20].

Zawawy et al. [21] present an interesting method that combines dynamic analysis of service logs with preliminary manual encoding of requirements information in goal trees. Their objective is to aid in root cause analysis of failures during corrective maintenance. Their method compares the events recorded in the logs with the goal trees describing expected behavior to locate the fault that is the root cause for a failure.

Chen et al. [22] describe a general framework that can be used to collect dynamic information to monitor SOA applications for a wide variety of mainte-nance and evolution tasks. They aim to integrate monitoring techniques into web service frameworks, so that the information for dynamic analysis will be trans-parently available for all applications using the framework.

Espinha et al. also use dynamic analysis to describe the runtime topology of a SOA application, by which they mean identifying which services are running, and how they depend and interact with one another. They provide an interesting analysis of the data a maintainer needs for different evolution scenarios. Their Serviz tool intercepts incoming requests to each service to capture the data they need for system visualizations [23].

The dynamic analysis approaches have many advantages. As can be seen from the examples we have cited, dynamic data can support striking visualizations to provide insight into the running system. Also dynamic data comes from the as-built system and thus, unlike models or documentation, is reliably up to date. However it can be difficult to take data from a running system at all the necessary points without encountering instrumentation, performance, and even confidentiality diffi-culties. Also, any dynamic data depends on what the system was doing at the moment when it was being observed; exceptional or rare behavior may be missed.

Static analysis methods, on the other hand, try to help a maintainer understand a SOA application without having to run it. They thus avoid the data collection problems of dynamic analysis, but with some costs.

One simple approach is to build on well-established search technologies to support SOA maintainers. The SOAMiner tool searches both text documentation and WSDLs, XML data schemas, and Business Process Execution Language (BPEL) code [24]. As well as conventional search, the tool also has a rule-based SOA Intel component that creates searchable abstractions from any XML structured inputs. The abstractions summarize relationships within the system and were defined based on the results of SOA comprehension case studies [25].

The problem with the search-based approach is that it largely leaves it up to the maintainer to formulate queries and understand the responses. The abstractions extracted from WSDLs, data descriptions, and BPEL can go only so far in providing high-level understanding of the system. However to go beyond the search-based approach seems to require additional inputs which may or may not be present for an existing SOA application.

For example Coffey et al. reverse engineer the WSDLs, data descriptions and BPELs into automatically generated concept maps which provide a convenient visualization of the SOA system. The intent is to then conduct interviews with system experts to annotate these maps into a more complete system description that would document it for future maintenance [26].

In another static approach, Kabzeva et al. present a very interesting method that focuses on managing the relationships in a large service network. However as well as the WSDL and BPEL inputs mentioned previously, they require information in modeling notations such as Business Process Model and Notation (BPMN) and Event-Driven Process Chains (EPC). If this information is present they can provide a relationships model that would seem to be very useful for maintenance tasks such as impact analysis [27].

Similarly Bauer et al. propose the use of a SOA repository with advanced analysis capabilities to identify relationships between the services and perform several important kinds of analysis, both to detect already existing problems (as-is-analyses), as well as problems that might occur due to future service changes (what-if-analyses). However they do not make clear how their repository would be populated with information and what manual inputs may be required [28].

So for static analysis at the service interaction level, there is a tradeoff between undemanding approaches such as search that leave a great deal to the user, and more sophisticated approaches with more complete results, but that require human data collection or accurate pre-existing system models.

2.4.3 Model Level Approaches to SOA Evolution

Modeling approaches can be used to support SOA evolution. Such approaches focus on the development of models to represent service oriented or enterprise architectures and the use of such models to guide their maintenance and evolution.

One such example is the OASIS Reference Model for Service Oriented Architecture [29], which is an abstract framework that guides reference architectures [30]. The ESARC—Enterprise Services Architecture Reference Cube [31] (Fig. 2.2) is more specific and completes these architectural standards in the context of EAM—Enterprise Architecture Management, and extends these architecture standards for services and cloud computing.

ESARC provides an abstract model to support the integration of business architectures with application architectures and implementation of service-based enterprise systems, and with the technology and operation architecture. ESARC is an original architecture reference model, which provides an integral view for main interweaved architecture types. ESARC abstracts from a concrete business scenario or technologies. The Open Group Architecture Framework provides the basic

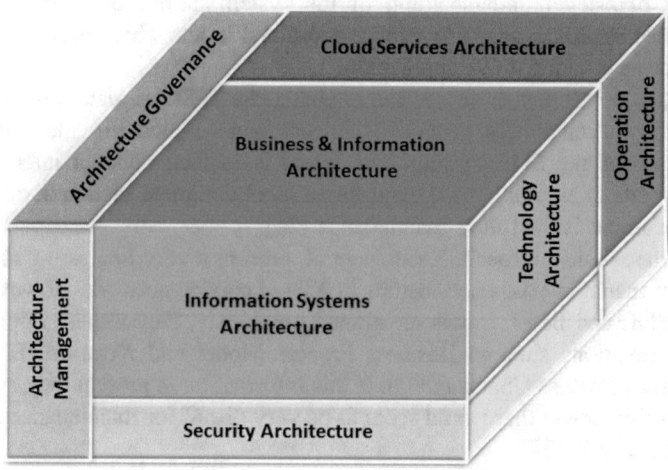

Fig. 2.2 ESARC—enterprise software architecture reference cube

blueprint and structure for the extended service-oriented enterprise software architecture domains like: Architecture Governance, Architecture Management, Business and Information Architecture, Information Systems Architecture, Technology Architecture, Operation Architecture, and Cloud Services Architecture. ESARC provides a coherent aid for the evolution of architectures by facilitating their examination, comparison, classification, quality evaluation and optimization of architectures.

Enterprise Architecture Management for Services Computing is a commonly preferred approach to organize, build and utilize distributed capabilities for Digital Transformation [32]. They provide flexibility and agility in business and IT systems. The development of such applications integrates the Internet of Things (IoT) [33], Web Services [34], REST Services [35], Cloud Computing [36] and Big Data [37], among other frameworks and methods, like architectural semantic support. Today's information systems span a broad range of domains including: intelligent mobility systems and services, intelligent energy support systems, smart personal health-care systems and services, intelligent transportation and logistics services, smart environmental systems and services, intelligent systems and software engineering, intelligent engineering and manufacturing. Microservices [38–40] and the Internet of Things [33] are examples of base technologies for the fast performing digital transformation. The Internet of Things enables a large number of physical devices to connect each other to perform wireless data communication and interaction using the Internet as a global communication environment.

Research reported in [41] focuses on extending the Enterprise Services Architecture Reference Cube (ESARC) by mechanisms for architectural integration and evolution to support adaptable information systems and architectural transformations for changing architectural models. ESARC is an extendable classification framework, which sets a conceptual baseline for digital architectural models.

ESARC makes it possible to verify, define and track the improvement path of different business and IT changes considering alternative business operating models, business functions and business processes, enterprise services and systems, their architectures and related technologies.

To integrate a huge amount of dynamically growing architectural descriptions for services, microservices, or the Internet of Things into consistent enterprise architectures is a considerable challenge. Further research focuses on integrating small EA descriptions for each relevant IoT object. EA-IoT-Mini-Description consists of partial EA-IoT-Data, partial EA-IoT-Models, and partial EA-IoT-Metamodels associated with main IoT objects like IoT-Resource, IoT-Device, and IoT-Software-Component [33]. This research addresses questions such as how can we federate these EA-IoT-Mini-Descriptions to a global EA model and information base by promoting a mixed automatic and collaborative decision process [42]. For the automatic part, model federation and transformation approaches are extended by introducing sematic-supported architectural representations, e.g. by using partial and federated ontologies and associated mapping rules—as universal enterprise architectural knowledge representation, which are combined with special inference mechanisms [43–46].

Metamodels can be used to define architecture model elements and their relationships within ESARC. These metamodels serve as an abstraction for architectural elements and relate them to architecture ontologies [47]. The OASIS Reference Model for SOA [29] is an abstract framework, which defines generic elements and their relationships for service-oriented architectures. Models and metamodels such as ESARC allow software maintainers to navigate the multidimensional space of service oriented and enterprise architectures and facilitate the development and use of semantic-supported navigation and intelligent inferences.

For years semantic technologies were said to revolutionize the web but for the time being the adoption rate is rather low. The basic idea of semantic web is to make the content of the web understandable for machines via the creation of semantic knowledge bases called ontologies. Semantic Web Services are typically extensions to conventional web services [48]. Semantic web services add extra semantic information in order to support automatic web service discovery, automatic web service invocation, automatic web service composition and interoperation [49]. Model Driven Architecture (MDA) uses UML as its preferred modeling language. Semantic models are extremely expressive when modeling structural knowledge. This facilitates modeling as well as maintenance of a model.

Salhofer et al. [50] presents an approach to apply the principles of Model Driven Architecture (MDA) combined with a semantic model. Model Driven Architecture focuses on the creation of models that should be turned into code automatically by code generators. The core idea is to create a model of a system that only represents its functionality but is not influenced by any technological platform. This model is called the Platform Independent Model (PIM). From the PIM, the Platform Specific Model (PSM) is generated. The PSM is then turned into source code by a code generator.

2.5   Emerging Trends

This section describes several emerging trends that can support the evolution of SOA systems.

2.5.1   Microservices and Design for Evolvability

A recent trend in many software application domains has been the shortening of software product delivery cycles. Companies have recognized that there is a strong commercial advantage to providing new features to customers ahead of their competitors. Software is often now delivered as a web application, perhaps combined with a client "app" automatically pushed to customer smartphones. In this environment the customer gets each new version transparently and there is no barrier to releasing new software daily or hourly. Terms used to describe this new software production model include *continuous deployment, continuous delivery* and *DevOps* since the roles of software developers and IT system operators become merged [51].

To support this model, new software engineering practices are being adopted such as small teams, tight communication between developers and other stakeholders, identical development and production environments, automatic build on commit, automatic testing on build, and automatic deployment on successful test. For these practices to work, the architecture of the application needs to be carefully planned.

Microservices is a name that has been given to an architectural style intended to work within these foreshortened delivery cycles [40]. The term is still relatively new and there is controversy about exactly what constitutes a microservices architecture. Recently at the 11th SEI Architecture Technology User Network (SATURN) Conference, a workshop characterized microservices as shown in Fig. 2.3 [38, 39].

The concept is that there will be small teams each responsible for a few small independent services. Teams will work at their own pace deploying new versions of services when they are ready, without having to coordinate versions. In our terminology, this is clearly a "design for evolvability" approach to SOA evolution. It remains to be seen if this approach will stand the test of time.

2.5.2   Knowledge-Based Support

Knowledge-based methods can support the evolution of increasingly complex SOA systems of the future. Such methods involve the use of knowledge representations to model SOA systems and reasoning strategies to support maintenance tasks and

Fig. 2.3 The microservices
architectural style

The SOA Architectural style, roughly consistent of these
constraints:

- Communication via messages

- Each service is independently deliverable

- Loosely coupled

Plus organizational constraints

- Decentralized design authority

- Architect is a coach

- Architecture is enforced through tooling

- Limited team size

In order to

- Sustain high delivery velocity by removing
  contention between teams and allowing rapid
  evolution (i.e. business agility and responding to
  change)

code comprehension. In particular, ontological modeling can support SOA evolution by representing both high-level business and architectural views of the whole application as well as lower level, code-focused views.

A commonly used ontology, the Open Group SOA ontology, can be extended to develop a *SOA Evolution Ontology* that better addresses software maintenance demands [30]. The Open Group's ontology describes business processes, services and their interfaces in a fairly abstract manner. The maintainer needs that description, but also needs to deal with concrete implementation details as may be found in design rationale, detailed interface specifications and in code. As an example of this approach, an ontology was developed to support semantic browsing and help a maintainer quickly acquire the information needed for a particular maintenance task [52]. A specialized semantic browser can be used to support the navigation of the large repositories of textual, semi-structured artifacts describing a SOA system. Textual artifacts include natural language design rationale, design and code documentation, semi-formal service interface specifications (e.g. WSDLs), BPEL orchestration code, etc. These artifacts are annotated through semantic labels that support discovery of the semantic relations between different artifacts.

Although little research has been reported on the development and use of knowledge-based methods for the maintenance and evolution of SOA systems there is literature on the application of semantic web techniques for maintaining traditional (non-SOA) software systems. The research reported in [52] focused on providing ontological support for software artifacts such as source code and documentation. In work reported by Witte et al. [53], customized ontologies were populated automatically from source code and documentation, and then queried to provide support for source code security analysis, for traceability links between source code and documentation and for architecture analysis. In work by Hyland-Wood et al. [54], an ontology was developed to describe the relationship between object-oriented software components.

Rastgoo et al. [55] is one of the few papers that take a knowledge-based approach to facilitate software engineering processes for SOA. The paper proposes automated generation of requirements ontologies using UML diagrams. The generated ontology considers the behavior and hierarchical relationship of services. Experimental results demonstrate the improvement of the proposed approach from perspectives, such as completeness and automatic generation of requirements ontology for SOA systems.

Knowledge-based support and ontological models can help address SOA evolution challenges to keep them in continuous service in the face of rapidly changing environments, continually emerging security risks, and a dynamic mix of partner organizations. Future trends will see the development of *ecosystems of ontologies* to describe increasingly complex SOA systems [41]. Consistent modeling approaches will need to emerge to bridge architectural levels and address the different concerns of business experts, developers and maintainers. The task of supporting the evolution of SOA systems will always be challenging, but such knowledge-based models could greatly ease the burden on software maintainers.

2.6 Concluding Remarks

In this chapter we described the challenges of software maintenance and evolution and examined various approaches specifically within the context of Services Oriented Architectures. We argue that SOA design principles such as loose coupling and service interfaces can facilitate SOA system evolvability. We described various approaches for SOA evolution support, which were classified by their level of granularity: service code level, service interaction level and model level approaches. We also presented emerging trends in supporting the maintenance and evolution of SOA systems, including microservices and knowledge-based support. Approaches such as the ones examined in this chapter can enhance the evolution of future SOA systems.

References

1. Lientz, B.P., Swanson, E.B., Tompkins, G.E.: Characteristics of application software maintenance. Commun. ACM **21**(6), 466–471 (1978)
2. Gold, N., Mohan, A., Knight, C., Munro, M.: Understanding service-oriented software. Softw. IEEE **21**(2), 71–77 (2004)
3. CanforaHarman, G., Di Penta, M.: New frontiers of reverse engineering. Future Softw. Eng. (2007) (IEEE Computer Society)

4. Lewis, G., Smith, D.B.: Service-oriented architecture and its implications for software maintenance and evolution. Frontiers Softw. Maint. (FoSM) (2008) (IEEE)
5. Kontogiannis, K.: Challenges and opportunities related to the design, deployment and, operation of Web Services. Front. Softw. Maint. (FoSM) (2008) (IEEE)
6. Lewis, G.A., Smith, D.B., Kontogiannis, K.: Proceedings of the Fourth International Workshop on a Research Agenda for Maintenance and Evolution of Service-Oriented Systems (MESOA 2010) (2011)
7. Parnas, D.L.: On the criteria to be used in decomposing systems into modules. Commun. ACM **15**(12), 1053–1058 (1972)
8. Koenemann, J., Robertson, S.P.: Expert problem solving strategies for program comprehension. In: Proceedings of the SIGCHI Conference on Human Factors in Computing Systems. ACM (1991)
9. Von Mayrhauser, A.: Program comprehension during software maintenance and evolution. Computer **28**(8), 44–55 (1995)
10. Erl, T.: SOA Principles of Service Design, vol. 37, pp. 71–75. Prentice Hall, Boston (2007)
11. Josuttis, N.M.: SOA in Practice: The Art of Distributed System Design. O'Reilly (2007). ISBN 0-596-52955-4
12. Christensen, E., Curbera, F., Meredith, G., Weerawarana, S.: Web services description language (WSDL) 1.1 (2001)
13. Borovskiy, V., Mueller, J., Schapranow, M., Zeier, A.: Ensuring service backwards compatibility with generic web services. In: Proceedings of the 2009 ICSE Workshop on Principles of Engineering Service Oriented Systems. IEEE Computer Society (2009)
14. Daigneau, R.: Service Design Patterns: Fundamental Design Solutions for SOAP/WSDL and Restful Web Services. Addison-Wesley (2011)
15. Scholl, H.J., Klischewski, R.: E-government integration and interoperability: framing the research agenda. Int. J. Publ. Adm. **30**(8–9), 889–920 (2007)
16. Glass, R.L.: Facts and Fallacies of Software Engineering. Addison Wesley (2003)
17. Hewitt, E.: Java SOA Cookbook. O'Reilly Media Inc. (2009)
18. Papazoglou, M.P., Andrikopoulos, V., Benbernou, S.: Managing evolving services. Softw. IEEE **28**(3), 49–55 (2011)
19. De Pauw, W., Lei, M., Pring, E., Villard, L., Arnold, M., Morar, J.F.: Web services navigator: visualizing the execution of web services. IBM Syst. J. **44**(4), 821–845 (2005)
20. Coffey, J., White, L., Wilde, N., Simmons, S.: Locating software features in a SOA composite application. In: 2010 IEEE 8th European Conference on Web Services (ECOWS). IEEE (2010)
21. Zawawy, H., Mylopoulos, J., Mankovskii, S.: Requirements-driven framework for root cause analysis in SOA environments. In: Proceedings of the Fourth International Workshop on a Research Agenda for Maintenance and Evolution of Service-Oriented Systems (MESOA 2010) (2011)
22. Chen, C., Zaidman, A., Gross, H.: A framework-based runtime monitoring approach for service-oriented software systems. In: Proceedings of the International Workshop on Quality Assurance for Service-Based Applications. ACM (2011)
23. Espinha, T., Zaidman, A., Gross, H.G.: Understanding the runtime topology of service-oriented systems. In: 2012 19th Working Conference on Reverse Engineering (WCRE), pp. 187–196. IEEE (2012)
24. Wilde, N., Leal, D., Goehring, G., Terry, C.: Enhanced search: an approach to the maintenance of services oriented architectures. In: Ninth International Conference on Software Engineering Advances (ICSEA 2014). Nice, France, 12–16 Oct 2014
25. El-Sheikh, E., Reichherzer, T., White, L., Wilde, N., Coffey, J., Bagui, S., et al.: Towards enhanced program comprehension for service oriented architecture (SOA) systems (2013)
26. Coffey, J.W., Reichherzer, T., Owsnick-Klewe, B., Wilde, N.: Automated concept map generation from service-oriented architecture artifacts, pp. 49–56 (2012)

27. Kabzeva, A., Götze, J., Lottermann, T., Müller, P.: Service relationships management for maintenance and evolution of service networks. In: The Eighth International Conference on Software Engineering Advances (ICSEA 2013) (2013)
28. Bauer, T., Buchwald, S., Tiedeken, J., Reichert, M.: A SOA repository with advanced analysis capabilities-improving the maintenance and flexibility of service-oriented applications (2015)
29. MacKenzie, C.M., Laskey, K., McCabe, F., Brown, P.F., Metz, R., Hamilton, B.A.: Reference model for service oriented architecture 1.0, p. 12. OASIS Standard (2006)
30. Open Group: Service-oriented architecture ontology (2010)
31. Zimmermann, A., Buckow, H., Gross, H., Nandico, O.F., Piller, G., Prott, K.: Capability diagnostics of enterprise service architectures using a dedicated software architecture reference model. In: 2011 IEEE International Conference on Services Computing (SCC). IEEE (2011)
32. Zimmermann, A., Schmidt, R., Sandkuhl, K., Jugel, D., Moehring, M., Wissotzki, M.: Enterprise architecture management for the internet of things. Lecture Notes in Informatics (2015), Dec 15, Boeblingen, Germany
33. Patel, P., Cassou, D.: Enabling high-level application development for the internet of things. J. Syst. Softw. **103**, 62–84 (2015)
34. Papazoglou, M.P., Web Services & SOA: Principles and Technology. Pearson—Prentice Hall (2012)
35. Ebert, J., Erl, T., Carlyle, B., Pautasso, C., Balasubramanian, R.: SOA with REST: Principles, Patterns & Constraints for Building Enterprise Solutions with REST. ACM SIGSOFT Software Engineering Notes, vol. 38(3), pp. 32–33 (2013)
36. Marinescu, D.C.: Cloud Computing: Theory and Practice. Newnes (2013)
37. Berman, J.J.: Principles of Big Data: Preparing, Sharing, and Analyzing Complex Information. Newnes (2013)
38. Microservices Workshop at SATURN 2015 [Internet]. Available from: https://saturnnetwork. wordpress.com/2015/05/07/microservices-workshop-at-saturn-2015
39. SATURN2015-Microservices-Workshop Key Outcomes [Internet]. Available from: https:// github.com/michaelkeeling/SATURN2015-Microservices-Workshop/blob/master/outcomes/ key-outcomes.md
40. Newman, S.: Building Microservices. O'Reilly Media, Inc. (2015)
41. Zimmermann, A., Gonen, B., Schmidt, R., El-Sheikh, E., Bagui, S., Wilde, N.: Adaptable enterprise architectures for software evolution of smart life ecosystems. In: 2014 IEEE 18th International Enterprise Distributed Object Computing Conference Workshops and Demonstrations (EDOCW). IEEE (2014)
42. Jugel, D., Schweda, C.M., Zimmermann, A.: Modeling decisions for collaborative enterprise architecture engineering. In: Advanced Information Systems Engineering Workshops. Springer (2015)
43. Breu, R., Agreiter, B., Farwick, M., Felderer, M., Hafner, M., Innerhofer-Oberperfler, F.: Living models-ten principles for change-driven software engineering. Int. J. Softw. Inform. **5** (1–2), 267–290 (2011)
44. Farwick, M., Pasquazzo, W., Breu, R., Schweda, C.M., Voges, K., Hanschke, I.: A meta-model for automated enterprise architecture model maintenance. In: 2012 IEEE 16th International Enterprise Distributed Object Computing Conference (EDOC). IEEE (2012)
45. Trojer, T., Farwick, M., Häusler, M., Breu, R.: Living modeling of IT architectures: challenges and solutions. In: Software, Services, and Systems, pp. 458–474. Springer (2015)
46. Khan, N.A.: Transformation of enterprise model to enterprise ontology (2011)
47. Zimmermann, A., Zimmermann, G.: Enterprise architecture ontology for services computing. In: Service Computation, pp. 64–9 (2012)
48. Alonso, G., Casati, F., Kuno, H., Machiraju, V.: Web Services. Springer (2004)
49. Martin, D., Paolucci, M., McIlraith, S., Burstein, M., McDermott, D., McGuinness, D., et al.: Bringing semantics to web services: the OWL-S approach. In: Semantic Web Services and Web Process Composition, pp. 26–42. Springer (2005)
50. Salhofer, P., Stadlhofer, B.: Semantic MDA for e-government service development. In: 2012 45th Hawaii International Conference on System Science (HICSS). IEEE (2012)

51. Wikipedia: DevOps [Internet] (2015). Available from: http://en.wikipedia.org/wiki/DevOps
52. Gonen, B., Fang, X., El-Sheikh, E., Bagui, S., Wilde, N., Zimmermann. A., et al.: Maintaining SOA Systems of the future—how can ontological modeling help? In: KEOD 2014—Proceedings of the International Conference on Knowledge Engineering and Ontology Development, Rome, Italy, 21–24 Oct 2014
53. Witte, R., Zhang, Y., Rilling, J.: Empowering software maintainers with semantic web technologies. In: The Semantic Web: Research and Applications, pp. 37–52. Springer (2007)
54. Hyland-Wood, D., Carrington, D., Kaplan, S.: Towards a software maintenance methodology using semantic web techniques and paradigmatic documentation modelling. Softw. IET **2**(4), 337–347 (2008)
55. Rastgoo, V., Hosseini, M., Kheirkhah, E.: Semantic web-based software engineering by automated requirements ontology generation in SOA. Int. J. Web Seman. Technol. **5**(2), 1 (2014)

Author Biographies

Norman Wilde is Nystul Chair and Professor of Computer Science at the University of West Florida. He received his Ph.D. in Mathematics and Operations Research from the Massachusetts Institute of Technology in 1971. His research interests are Software Engineering, Software Maintenance/Evolution, Services Oriented Architectures and Cybersecurity.

Bilal Gonen is an Assistant Professor of Information Technology at the University of Cincinnati. He received his Ph.D. in Computer Science and Engineering from University of Nevada, Reno, in 2011. His research interests are Software Engineering, Services Oriented Architectures, Computer Networks, Complex Networks, Social Network Analysis, Semantic Web, Algorithms, Machine Learning.

Eman El-Sheikh is Professor of Computer Science and Director of the Center for Cybersecurity at the University of West Florida. She received her Ph.D. in Computer Science from Michigan State University in 2001. Her research interests include Artificial Intelligence, Machine Learning, Intelligent Systems, Cybersecurity, Software Maintenance and Evolution, and Services Oriented Architectures.

Alfred Zimmermann is Professor of Computer Science at Reutlingen University and Research Director of the Herman Hollerith Center for Services Computing Boeblingen, Germany. His research is focused on Digital Transformation and Digital Enterprise Architecture in close relationship with Services and Cloud Computing. He graduated in Medical informatics at the University of Heidelberg and got his Ph.D. in Informatics from the University of Stuttgart, Germany.

Chapter 3
Flexible and Maintainable Service-Oriented Architectures with Resource-Oriented Web Services

Michael Gebhart, Pascal Giessler and Sebastian Abeck

Abstract The implementation of service-oriented architectures is mostly driven by the motivation to create a flexible and maintainable IT. Whether this goal can be achieved or not strongly depends on the design quality of the services. For that reason, the services within a service-oriented architecture have to be created with care. In the past, several quality attributes and quality indicators were identified that provide information about the design quality of a service. These quality indicators were described with focus on method-driven services based on SOAP. However, today, services are often designed in a resource-oriented way using REST or similar approaches to enable technology-independent interactions. For that reason, this chapter maps the existing quality attributes and quality indicators onto resource-oriented web services. As result, architects and developers get a toolset to design resource-oriented web services in a service-oriented architecture systematically in a quality-oriented manner. The quality indicators are illustrated by means of a resource-oriented web service in the context of a service-oriented SmartCampus system developed at the Karlsruhe Institute of Technology. The scenario shows that the application of the quality indicators limits the design scope and accelerates making design decisions.

3.1 Introduction

Today, the motivation behind the implementation of service-oriented architectures is mostly the creation of a flexible and maintainable IT. The quality of the resulting architecture is strongly influenced by the quality of its building blocks which are represented by services. For that reason, services also have to be designed in a quality-oriented manner. In the last years, several quality attributes were identified

M. Gebhart (✉) · P. Giessler
iteratec GmbH, Stuttgart, Germany
e-mail: michael.gebhart@iteratec.de

S. Abeck
Karlsruhe Institute of Technology, Karlsruhe, Germany

© Springer International Publishing Switzerland 2016
E. El-Sheikh et al. (eds.), *Emerging Trends in the Evolution of Service-Oriented and Enterprise Architectures*, Intelligent Systems Reference Library 111,
DOI 10.1007/978-3-319-40564-3_3

that are considered as important for services in service-oriented architectures. Examples are loose coupling, autonomy, and discoverability introduced by Erl [1]. As quality attributes describe only abstract concepts, the evaluation of services regarding their quality requires the refinement of quality attributes into measurable elements, the so-called quality indicators. Previous work by Gebhart and Abeck [2] introduced quality indicators for the most widespread quality attributes. These indicators refer to the interface of services and their internal logic. To evaluate concrete web service implementation artifacts regarding these quality indicators, a mapping of these conceptual indicators onto concrete technology is necessary. For this purpose, in [3], Gebhart mapped the indicators onto the Web Service Description Language (WSDL) as language to describe the web service interface and onto the Service Component Architecture (SCA) to model the internal logic. Based on this mapping, the quality indicators can be partially automatically measured.

However, due to their lightweight and technology-independence, more and more resource-oriented approaches, such as REpresentation State Transfer (REST) introduced by Fielding [4], based on Hypertext Transfer Protocol (HTTP) are chosen to implement web services in service-oriented architectures. Resource-oriented approaches differ from the method-oriented design represented by for instance SOAP and WSDL. In the past, existing work described quality attributes only in an abstract manner [2] or mapped them onto SOAP-based web services [3]. Thus, for architects and developers it is not obvious, whether developed resource-oriented web services consider the quality attributes correctly. Therefore, it is necessary to map important quality attributes for service-oriented architectures and their indicators onto service designs with resource-orientation in mind.

This article maps the quality indicators derived by Gebhart and Abeck [2] from widespread literature onto the concepts of resource-oriented web services. For that purpose, this article shows for each quality indicator, which aspects of a resource-oriented web service, such as the Unified Resource Identifier (URI) or the design of its parameters, influence the quality indicator. Furthermore, the desired characteristic is determined. This will help architects and developers to understand how to design resource-oriented web services in a way that the design supports the flexibility and maintainability of the service-oriented architecture. In this chapter, the focus is on resource-orientation in general over HTTP and not RESTful web services in special. According to Fielding in [4], a truly RESTful web service requires the application of Hypermedia As The Engine Of Application State (HATEOAS). However, this methodology is mostly not applied in industry due to its implementation complexity. For that reason, we focus on resource-orientation without hypermedia. I.e., we do not consider RESTful but resource-oriented web services. According to the Richardson Maturity Model (RMM), this kind of web service is positioned on the second maturity level [5].

To illustrate the quality indicators and their mapping onto resource-oriented web services, the AccessibilityInfoService of the service-oriented SmartCampus system at the Karlsruhe Institute of Technology (KIT) is designed and developed considering the quality indicators. The SmartCampus supports members, guests, and students at the KIT in their daily campus life. For example, they can determine routes to certain rooms or can find free workplaces. The AccessibilityInfoService is a new web service for people with disability, developed in cooperation with the study centre for visually impaired (SZS). The service provides detailed information about the accessibility of building and lecture rooms in a barrier-free way according to the Web Content Accessibility Guidelines (WCAG) in version 2.0 [6]. The application of the quality indicators by means of this web service will show how the quality indicators support architects and developers during the design and development of a resource-oriented web service.

This book chapter is structured as follows: Sect. 3.2 gives a brief introduction into the fundamentals, such as SOAP, REST, and the used quality model. The scenario, i.e., the SmartCampus and the AccessibilityInfoService are described in Sect. 3.3. Section 3.4 introduces the quality indicators for service-oriented architectures and maps them onto resource-oriented web services. Section 3.5 concludes this book chapter and presents an outlook on future research work.

3.2 Fundamentals

The following section forms the basis on which the described concepts in this book chapter build on. For interface design, there are two types of service interfaces: (1) proprietary and platform-dependent interfaces and (2) standardized and platform-independent interfaces [7]. Due to our focus on platform-independent service interfaces, we only consider standardized interfaces that are frequently used for web services. For web services, there are two widespread distinctive types: (1) SOAP-based web services and (2) web services based on Representation State Transfer (REST) with increasing popularity [7, 8]. Each of them is presented by a subsection. After describing and illustrating the two web service types, a quality model derived from the ISO/IEC 25010:2011 [9] is introduced to define the terms for service design quality.

3.2.1 SOAP

SOAP is a wire protocol, which defines the format and structure of the transmitting data [10, 11]. For transmission of SOAP messages, the application layer protocol HTTP (Hypertext Transport Protocol) is typically used [12]. The Web Service Description Language (WSDL) describes the interface of SOAP-based web services and the exchanged messages. By using WSDL, three different aspects of a web service can be defined: (1) the exposed operations of the service, (2) the input and output parameters for these operations, and (3) the technology binding by which a client can communicate with the web service [7]. Compared to resource-oriented web services, this web service type can be denoted as a method-oriented web service due to its exposed operation declarations.

3.2.2 REST

REST is a hybrid architectural style as result of combining different network-based architectural styles together with another constraint for the uniform interface [4]. An architectural style can be seen as a crosscutting concept for designing software architectures, which can be grouped in different categories. Garlan and Shaw [13] define an architectural style as a "vocabulary of components and connectors that can be used in instances of that style, together with a set of constraints on how they can be combined" [13, p. 6].

For the design of REST, Fielding [4] has identified the following four key characteristics, which were proved as responsible for the success of the World Wide Web (WWW) [14]: (1) Low entry-barrier, (2) extensibility, (3) distributed hypermedia, and (4) internet-scale. REST ensures these characteristics through the following six constraints:

- Client and Server: A client component sends a request to a server component for executing a remote operation. It is incumbent upon the server component to perform or reject the request [4].
- Statelessness: Each request from client to server has to contain all necessary information to perform the request, which leads to the following advantage: "There is no need for the server to maintain an awareness of the client state beyond the current request" [4, p. 199].
- Layered Architecture: A layered client-server architecture enables the application of the Separation of Concerns (SoC) principle and the opportunity to add features like load balancing or caching mechanisms to multiple layers [4, 15].
- Caching: This constraint allows a client to match its request to a previous response from the server with the result that no request has to be transmitted over the network [14].
- Code on Demand: With the usage of Code on Demand, additional programming logic can be requested from the server that is needed for processing information received from the server [14].
- Uniform Interface: The term "uniform interface" (hereinafter the resource-oriented interface) can be seen as an umbrella term, since it can be decomposed into four sub-constraints [14]: (1) identification of resources, (2) manipulation of resources through representations, (3) self-descriptive messages, and (4) hypermedia.

Typically, web services based on REST will be used in combination with the application layer protocol HTTP [7, 12]. But, also other protocols can be used such as CoAP [14, 16]. Through the usage of hypermedia and semantic markups, no further information is necessary to interact with this type of web services [5, 9].

3.2.3   Quality Model

The following quality model is based on the ISO/IEC 25010:2011 [9] and builds the foundation for the qualitative perspective in this book chapter. Besides the definition of the quality terms, also the relationship between them will be defined. Together, they form the basis for evaluating service design quality in the upcoming sections.

A quality characteristic represents a "category of software quality attributes that bears on software quality" and can be refined into multiple sub-characteristics over several levels according to the ISO/IEC 25010:2011 [9]. Finally, there are quality attributes on which an "inherent property or characteristic of an entity can be distinguished quantitatively or qualitatively by human or automated means" [9]. The quality attributes thus permit a statement about the manifestation of one or more quality characteristics of the service under investigation. To get a hint about the manifestation of a quality attribute, so-called quality indicators can be used. The quality indicators are represented by quality metrics that allow a "measurement of the degrees to which software has given quality attributes" [17]. Quality indicators can be derived from design patterns, best practices, conventions, or similar quality influencing concepts. On the other hand, the quality criteria defines the threshold "for satisfaction and the corresponding measures" [9]. For illustration purpose, Fig. 3.1 shows the quality model.

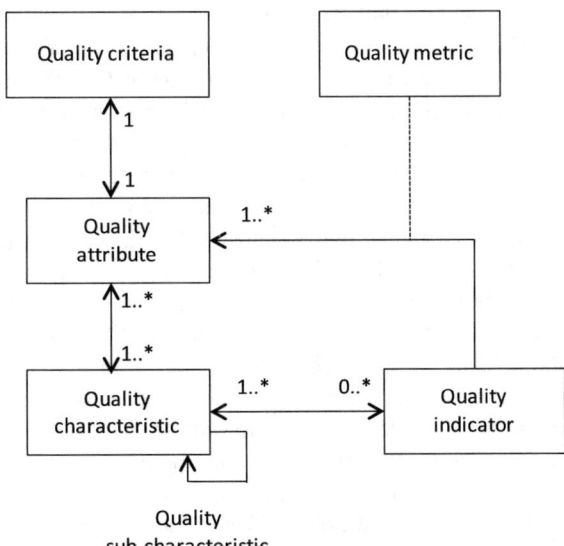

Fig. 3.1 Quality model applied in this chapter based on ISO/IEC 25010:2011

In [2], Gebhart derived four central quality attributes for services from existing work: unique categorization, loose coupling, discoverability, and autonomy. As basis, the work of Erl [1], Cohen [18], Perepletchikov et al. [19–21], Hirzalla et al. [22], and Choi et al. [23] were used. These quality attributes influence the flexibility and maintainability of service-oriented architectures. Existing work describes these quality attributes mostly only abstractly and textually. Therefore, architects and developers cannot apply them without additional interpretation effort. For that reason, they are not sufficient to design services systematically in a quality-oriented way. Instead, a quality attribute has to be refined into quality indicators that can be evaluated on concrete design or implementation artifacts. In [2], Gebhart introduced quality indicators for services in general. These quality indicators constitute the basis for this book chapter.

3.3 Scenario

This section provides a detailed description of the AccessibilityInfoService, a web service as part of the service-oriented SmartCampus system at the KIT, at which the designed and developed quality indicators were applied. The application proves our assumption that they limit the design scope and accelerate making design decision with quality in mind. For a generally valid statement, further software projects have to be realized while considering our provided set of quality indicators. Such an explorative study is outside of the scope of this chapter. However, this chapter provides the necessary information to accomplish such a study. The provided scenario shall therefore not act as an evidence of our quality indicators rather than it will be used to illustrate our identified quality indicators. In the following, the scope of the scenario will be described in detail.

The SmartCampus system provides services for several platforms, in particular for mobile devices, to support the daily life at the university. Today, there are several services, such as the ParticipationService for system consenting [24], the CampusGuide service for navigating on the campus, and the CompetenceService for searching competences with an ontology-based approach.

The AccessibilityInfoService is a new web service for people with disability, which is currently under development in cooperation with the study centre for visually impaired (SZS). The service provides detailed information about the accessibility of buildings and lecture rooms in a barrier-free way according to the Web Content Accessibility Guidelines (WCAG) in version 2.0 [6]. For example, there is information about main and side entrance of buildings regarding the existence of stairs, amount of stair steps, existence of a ramp function, and automatic door openers. Besides the retrieval of available information, there is also a notification system that notifies the people about risks in the near surroundings or requested buildings and lecture rooms.

From a technical point of view, the AccessibilityInfoService was designed and developed with modern state of the art technologies, such as AngularJS and Spring.

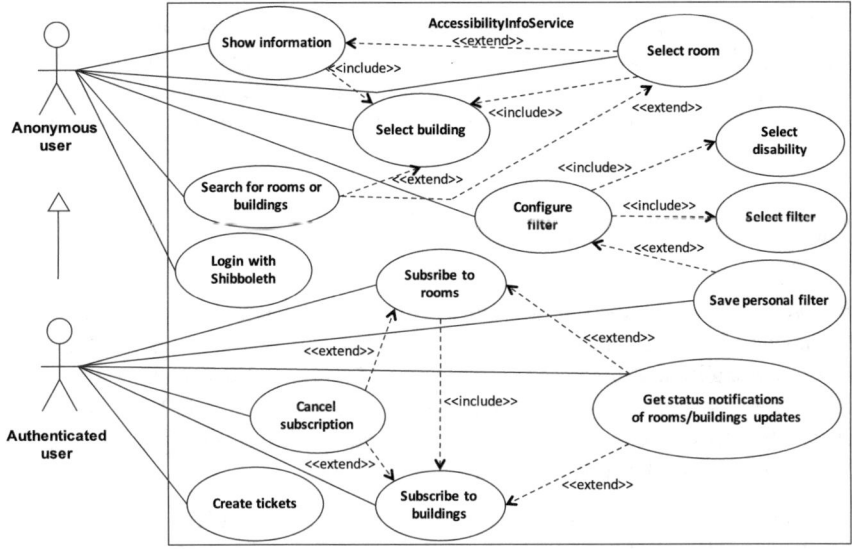

Fig. 3.2 Use cases covered by AccessibilityInfoService

For the service interface, a resource-oriented approach was chosen similar to REST with one exception: no hypermedia was used for state transitions according to the principle of HATEOAS. In the transmitted representations, there are no hyperlinks for further proceeding in form of new interactions with the service. Instead of this, the logic for further processing was implemented directly in JavaScript. Therefore, the web service is not RESTful according to the constraints by Fielding [4], but rather a resource-oriented web service. Compared to the maturity model by Richardson, the web service can be positioned on the second level [5].

Figure 3.2 illustrates the AccessibilityInfoService as part of the SmartCampus system at the KIT in form of a use case diagram from the perspective of people with disability. We differentiate between a user with and without authentication (anonymous) on the AccessibilityInfoService.

3.4   Quality Indicators for Resource-Oriented Web Services

This section introduces the quality attributes and the technology-independent quality indicators that have been identified as important for services in service-oriented architectures. Afterwards, each of the technology-independent quality indicators is mapped onto aspects of resource-oriented web services. This enables architects and developers to design resource-oriented web services systematically in a quality-oriented manner. Furthermore, by application of these quality indicators, existing resource-oriented web services can be evaluated regarding the introduced quality attributes to identify potential for improvement. The mapping of quality indicators onto aspects of resource-oriented web services is illustrated by means of the AccessibilityInfoService as introduced in Sect. 3.3.

3.4.1 Unique Categorization

According to Erl [1], the unique categorization in the context of service-orientation is comparable to the concept of cohesion in object-orientation. In general, it means that functionality that belongs together should be grouped into one service. On the other side, functionality with no focus on related aspects should be separated into several services. According to Gebhart and Abeck [2], this quality attribute can be broken down into the following quality indicators:

Separation of Business-Related and Technical Functionality
To increase the maintainability of services, business-related and technical functionality should be separated into different services because these two kinds of functionality change in different time intervals.

In the context of resource-oriented web services, one web service focuses on one resource that can be either business-related or technical. In case of business-relation, the covered resource reflects an entity of the domain model. According to this quality indicator, all operations of the web service, represented by HTTP methods, should relate to the considered resource. If technical functionality is required as part of a business-related resource, this functionality should be part of a separate web service representing a more technical resource. For example:

```
GET/users/bfc6cacf9bad9a02c87c3061a491b11b
```

This request returns the user with the md5-generated ID bfc6-cacf9bad9a02c87c3061a491b11b. To get the information whether the user is currently logged in, what represents a more technical functionality, the following way is a bad style as it is more method-oriented than resource-oriented:

```
GET/users/bfc6cacf9bad9a02c87c3061a491b11b/loggedIn
```

Instead, for the login information a separate resource-oriented web service should be created that allows querying login information for a specific user ID.

```
GET/logins/bfc6cacf9bad9a02c87c3061a491b11b
```

Separation of Agnostic and non-Agnostic Functionality
To increase the reusability of services, agnostic functionality should be separated from non-agnostic functionality. Agnostic functionality represents generic functionality that can be used in several business contexts. A typical example for an agnostic functionality is a user query with a certain user ID.

A resource-oriented web service typically provides information that relates to a resource representing a functional entity of the domain model. As the provided HTTP methods enable Create, Read, Update, and Delete (CRUD) operations that can be used in several contexts, the web service provides only agnostic functionality. For specific functionality, an explicit resource should be created. For example, if a web service is expected to provide functionality to subscribe to rooms or buildings, a separate resource for subscriptions should be created. This means that the following request is a bad style to subscribe to rooms with md5-based ID d8ce37c2d80891095497a73e37432d56.

`POST /rooms/d8ce37c2d80891095497a73e37432d56/subscribe`

Instead, the following resource is recommended:

`POST /subscriptions`

As payload, the service consumer sends the ID of the room, i.e., in this example d8ce37c2d80891095497a73e37432d56. Sometimes, especially behind non-agnostic functionality there are complex and long-running business processes. In this case, the quality indicator of asynchronicity as part of the loose coupling quality attribute should be considered.

Data Superiority
The statement behind data superiority is that there is only one service that is responsible for the management of a certain business entity, such as customers or invoices. In this context, management means the creation, reading, updating, and deletion (CRUD) of the business entity. For that reason, it is comparable to the Separation of Concerns (SoC) principle. The data superiority increases the maintainability since all managed functionality is collected in one central place. If the data schemas are changed, only one service has to be adapted.

For resource-oriented web services, the data superiority means: If there is a building service which provides information about buildings, then no further service with similar functionality for buildings should exist. Thus, if a resource-oriented web services manages only the covered resource, then the web service automatically fulfills this quality indicator.

Common Business Entity Usage
According to this quality indicator, a service should provide functionality that uses only common business entities. This means that all provided functionality uses either the same business entity or dependent ones. A business entity "A" depends on another one "B" if the entity "A" cannot exist without "B". An example is the business entity building with a contained address considered as an explicit business entity. A building service should focus on a building but as an address cannot exist without the superordinate building, the building service can also work with addresses. An explicit address service is not necessary. In domain models that are described using the Unified Modeling Language (UML), this dependency would be described by means of compositions.

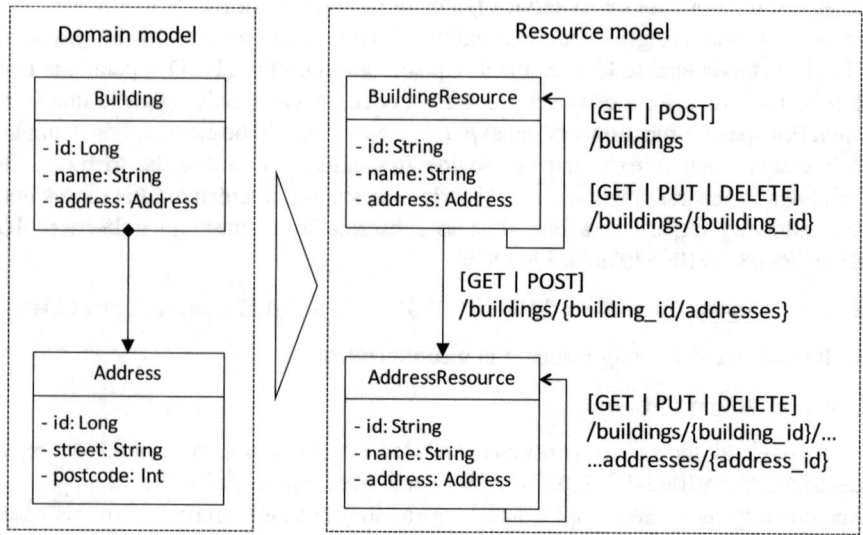

Fig. 3.3 Mapping composition from domain model onto resource model

For resource-oriented web services, this quality indicator means that one web service should only consider one resource and its composing resources that are often called sub-resources (Fig. 3.3).

3.4.2 Loose Coupling

Intention of a loose coupling between services is the reduction of its dependencies. This quality attribute represents one of the most widespread ones since it is often mentioned as an important success factor for service-oriented architecture projects. A loose coupling between services promotes the scalability, fault tolerance, flexibility, and maintainability of the entire service-oriented architecture. The following quality indicators provide architects and developers hints about the current degree of coupling.

Asynchronicity
To reduce the dependencies, in case of long-running operations, asynchronous operation calls are expected to be supported. This means that the service consumer invokes the service provider, who returns the result proactively by calling the service consumer. This enables that the service consumer does not have to wait until the operation has been finished. Asynchronicity decouples the service consumer from the service provider during the execution. As result, for example, the service consumer is not required to be active and correctly running during the entire execution time.

For that purpose, in resource-oriented web services, a POST request is sent to initiate the long-running operation.

```
POST /subscriptions
```

As payload, a Callback Resource Identifier should be provided by means of a URL that can be called by the operation to initiate the callback. In case of long-running operations, the POST request returns the ID of the process representing the operation execution. By this means, it is possible to query the current state of the operation or to terminate it.

```
GET /subscriptions/d8ce37c2d80891095497a73e37432d56
```

This request returns a representation of the operation with the process ID d8ce37c2d80891095497a73e37432d56.

As alternative to the Callback Resource Identifier, it would be possible to use polling to recognize when the operation is finished. However, especially in mobile environments, this solution is not recommended due to declining battery life. Another option is push services, which notify service consumer over well-defined interfaces about the process status. An interaction from client-side is not necessary.

Common Data Types Complexity
If service providers and consumers use the same data types, their coupling increases as changing the data types within one of the parties results in adapting the other party. For that reason, service consumer and service provider should not use the same data types in case of complex ones. The only shared data types should be the primitive ones, such as string or integer.

To consider this quality indicator in resource-oriented web services, the implementation has to be considered: When using JSON as representation data format, only strings are exchanged. Therefore, in this case, it depends on the internally used data types: If service providers and consumers use the same complex data types internally and create serialized representations of these data types to exchange data, then this indicator is not fulfilled. However, if service provider and consumer use similar data types contained in different packages or the JSON string is created manually, this indicator is fulfilled.

Abstraction
Another aspect that supports the loose coupling is the abstraction of operations and parameters. This means that both the operations and the expected parameters should hide technical details. For example, if the functionality of a service is the storage of users in the database, the service provider should only provide the abstract information that users are stored. The usage of a database should be hidden. Also in the context of parameters, only functional information should be expected. Implementation details, such as database credentials, should not be expected parameters. This enables to change the underlying implementation and the backend without changing the service consumer.

Table 3.1 HTTP method characteristics

HTTP method	Safe	Idempotent
POST	No	No
GET	Yes	Yes
PUT	No	Yes
DELETE	No	Yes

Regarding operations in resource-oriented web services, the URLs have to be considered as the operations themselves are preset by HTTP (GET, POST, PUT, and DELETE). A typical non-abstract URL is:

```
GET /buildingsInDatabase
```

Technical information, such as the data store should be avoided in the URL. Regarding the parameters, it is very similar to the method-oriented approaches as it depends on the implementation: The transferred data, e.g., described with JSON should not provide non-abstract information. For example, the following JSON object to create a new customer should be avoided:

```
POST /users
{

"lastname": "Doe",
"firstname": "John",
"db_username": "database-user1",
"db_password": "database-pw1"

}
```

Compensation

Aservice should provide compensating functionality in case of state-changing functionality. For example, if a service provides functionality to change the state of a business entity, such as users, it should also provide functionality to compensate this step. This is especially needed in transactional contexts.

For resource-oriented web services, this means that on the one hand the HTTP methods have to be correctly applied as described in Table 3.1.

And on the other hand, POST and DELETE always have to be available together as they are the compensating ones for each other. PUT is for updating an entry and if it is idempotent it is also self-compensating.

3.4.3 Discoverability

Reusing existing functionality is another important aspect in the context of service-oriented architectures. Certain functionality should only be available once and thus should not be implemented twice. For that reason, it is necessary that existing service are discoverable. The discoverability can be broken down into the following three quality indicators:

Functional Naming

To discover existing functionality, it is necessary that all externally visible parts of a service are functionally named. Compared to the abstraction introduced as quality indicator for the loose coupling, the functional naming is not limited to the avoidance of technical details. Instead the focus is on functional naming in general. For example, a service that is responsible for the management of buildings should be named as buildings service and not simply service A or service B. This has to be considered for all externally visible parts.

In resource-oriented web services, the operations themselves are preset by HTTP. Therefore, this part is fixed and cannot be changed. However, the URLs should be functionally named. A URL like the following should be avoided:

```
/A/d8ce37c2d80891095497a73e37432d56
```

Instead, the following URL should be chosen.

```
/buildings/d8ce37c2d80891095497a73e37432d56
```

Furthermore, the representations, i.e., the exchanged data, in case of JSON especially the attributes, have to be functionally named.

```
{

"I": 5331,
"BN": "01.13",
"N": "Mensa new building",
"ADD": "Straße am Forum 4",
"LA": 49.0118,
"LO": 8.41688,
"EMGNUM": null,
"IDK": false,
"MISC": null,
"CATID": 27053

}
```

Should be replaced by:

```
{

"id": 5331,
"buildingNumber": "01.13",
"name": "Mensa new building",
"address": "Straße am Forum 4",
"lat": 49.0118,
"lon": 8.41688,
"emergencyNumber": null,
```

```
"infodesk": false,
"miscellaneous": null,
"categoryId": 27053

}
```

Naming Convention Compliance

Both in literature and in companies there are naming conventions that should be considered when designing a service. For example, operations are expected to include verbs and nouns. Furthermore, capitalization rules should be correctly applied and there is no mixture of singular or plural. This helps architects and developers to find appropriate existing functionality.

Transferred to resource-oriented web services, this is very similar to the functional naming. Also in this case, the URLs and the attributes in data objects have to follow naming conventions. For example bad URLs could be:

```
/BUILDING/d8ce37c2d80891095497a73e37432d56
/ROOMS/c779b524addc6effc9334eea029208ca
```

Instead, the following convention should be chosen:

```
/buildings/d8ce37c2d80891095497a73e37432d56
/rooms/c779b524addc6effc9334eea029208ca
```

Information Content

The more information the service provides to potential service consumers the easier the appropriate functionality can be found. For that reason, the extent of information content, such as documentation of the service interface, indicates the discoverability of a service.

For resource-oriented web services, there is no formal description available as it is for web services based on SOAP and WSDL. For that reason, as much information as possible should be documented: The resources, available methods, attributes of exchanged data. Especially for the documentation of attributes, the Application Level Protocol Semantics (ALPS) can be chosen to provide the semantics for the exchanged data [25].

3.4.4 Autonomy

The autonomy represents the independency from one service to other services, i.e., to what extent a service can be used without other ones. To get a hint regarding the degree of autonomy of a service, architects and developers can apply the following quality indicators.

Service Dependency
Functionality that requires other services decreases the autonomy of the considered service since it requires the presence of other services for operating. For that reason, the direct dependencies between one service and other services give a hint about its autonomy.

In the context of resource-oriented web services, the implementation has to be considered to measure the number of dependencies to other services.

Functional Overlap
Furthermore, the autonomy is decreased if the provided functionality overlaps with the functionality of other services. In this case, the service can only be used in combination with functionality of other services what reduces its autonomous usage.

Similar to service dependency, this quality indicator has no REST-specific characteristic. The implementation has to be considered to evaluate the functional overlap with other services.

3.5 Conclusion and Outlook

In the past, mostly web services based on the method-oriented approaches, such as SOAP, have been developed. However, due to its lightweight, today, more and more resource-oriented web services based on HTTP are developed. This kind of web services can also be applied even though if the service-oriented architecture is implemented using certain trends, such as microservices [26]. As the design of these web services strongly influences the quality of the resulting service-oriented architecture, the web services have to be designed with care.

For that reason, quality attributes with focus on design and implementation aspects of services in service-oriented architectures have been described. As these quality attributes are mostly only described textually, Gebhart and Abeck [2] broke them down into measurable quality indicators. However, these quality indicators are still technology-independent. For that reason, in the past, they have been mapped onto method-oriented web services, such as the ones based on SOAP. In this chapter, the outstanding mapping onto resource-oriented web services was described.

To illustrate the mapping, the quality indicators were applied on a real-world scenario, the AccessibilityInfoService developed at the Karlsruhe Institute of Technology (KIT). The AccessibilityInfoService is a web service developed for people with disability in cooperation with the study centre for visually impaired (SZS) as part of the SmartCampus system. The application showed that the quality indicators give valuable hints about how to design the web services so that the desired quality attributes, such as loose coupling, are supported.

By means of the mapping onto resource-oriented web services, architects and developers get a tool to design and implement resource-oriented web services based on HTTP systematically in a quality-oriented manner. Furthermore, by means of the quality indicators, it is possible to evaluate existing web services to identify potential for improvement. For that reason, the present work supports the creation of service-oriented architectures that fulfill the goals initially associated with this kind of architecture: A highly flexible and maintainable IT architecture.

For the future, we plan to support the automatic evaluation of web services regarding the described quality indicators. For that purpose, we are working on an open source tool, the QA82 Analyzer [27]. As the evaluation regarding the quality indicators mostly requires expert knowledge, this tool supports an evaluation mechanism, where automatic evaluations are combined with manual expert knowledge. We call this approach "hybrid quality analysis". The tool allows the creation of further metrics and can be integrated into existing tool chains.

References

1. Erl, T: SOA: Principles of Service Design. Prentice Hall, Upper Saddle River (2007). ISBN 978-0132344821
2. Gebhart, M., Abeck, S.: Metrics for evaluating service designs BASED on SoaML. Int. J. Adv. Softw. 4(1&2), 61–75 (2011)
3. Gebhart, M.: Query-based static analysis of web services in service-oriented architectures. Int. J. Adv. Softw. 7(1&2), 136–147 (2014)
4. Fielding, R.: Architectural Styles and The Design of Network-Based Software Architectures. University of California, Irvine (2000)
5. Webber, J., Parastatidis, S., Robinson, I.: REST in practice: Hypermedia and Systems Architecture. O'Reilly Media Inc (2010)
6. W3C: Web Content Accessibility Guidelines (WCAG) 2.0 (2008)
7. Dikmans, L., Luttikhuizen, R.: SOA Made Simple. PACKT Publishing (2012). ISBN 9781849684163
8. Mason, R.: How REST Replaces SOAP on the Web: What it means to you. http://www.infoq.com/articles/rest-soap (2011)
9. ISO/IEC 25010: Systems and software engineering—Systems and software Quality Requirements and Evaluation (SQuaRE)—System and software quality models (2011)
10. ISO/IEC 40210: Information technology—W3C SOAP Version 1.2 Part 1: Messaging Framework (2011)
11. ISO/IEC 42020: Information technology—W3C SOAP Version 1.2 Part 2: Adjuncts (2011)
12. IETF RFC 2616: Hypertext Transfer Protocol—HTTP/1.1 (1999)
13. Garlan, D., Shaw, M.: An introduction to software architecture. Pittsburgh, PA (1994)
14. Richardson, L., Amundsen, M., Sam, R.: RESTful Web APIs. O'Reilly & Associates (2013). ISBN 978-1449358069
15. Evans, E.: Domain-Driven Design: Tacking Complexity In the Heart of Software. Addison-Wesley Longman Publishing Co, Boston (2003). ISBN 0321125215
16. IETF RFC 7252: The Constrained Application Protocol (CoAP) (2014)
17. Summers, B.L.: Software Engineering Reviews and Audits. CRC Press, Boca Raton (2011)
18. Cohen, S.: Ontology and taxonomy of services in a service-oriented architecture. Microsoft Archit. J. (2007)

19. Perepletchikov, M., Ryan, C., Frampton, K., Schmidt, H.: Formalising service-oriented design. J. Softw. **3**, 1–14 (2008)
20. Perepletchikov, M., Ryan, C., Frampton, K., Schmidt, H.: Cohesion metrics for predicting maintainability of service-oriented software. In: Seventh International Conference on Quality Software (QSIC) (2007)
21. Perepletchikov, M., Ryan, C., Frampton, K., Tari, Z.: Coupling metrics for predicting maintainability in service-Oriented design. In: Australian Software Engineering Conference (ASWEC) (2007)
22. Hirzalla, M., Cleland-Huang, J., Arsanjani, A.: A metrics suite for evaluating flexibility and complexity in service oriented architecture. In: ICSOC (2008)
23. Choi, S.W., Kimi, S.D.: A quality model for evaluating reusability of services in soa. In: 10th IEEE Conference on E-Commerce Technology and the Fifth Conference on Enterprise Computing, E-Commerce and E-Services (2008)
24. Gebhart, M., Giessler, P., Burkhardt, P., Abeck, S.: Quality-oriented requirements engineering for agile development of restful participation service. Ninth International Conference on Software Engineering Advances, pp. 69–74 (2014)
25. Amundsen, M., Richardson, L., Foster, M. W.: Application-Level Profile Semantics (ALPS). http://alps.io/spec/ (2014). Last-accessed 17 May 2015
26. Newman, S.: Building Microservices. O'Reilly (2015)
27. QA82: QA82 Analyzer. http://www.qa82.org

Author Biographies

Michael Gebhart is Senior IT Management Consultant at iteratec GmbH in Stuttgart, Germany. He did his Ph.D. in the context of quality-oriented design of services in service-oriented architectures. Michael Gebhart is author of numerous articles, speaker at conferences, lecturer at universities, and researcher in the areas of service-oriented architectures, web applications, and quality analysis of software.

Pascal Giessler is a software engineer, author and researcher working at iteratec GmbH in Stuttgart, Germany. He holds a master degree in computer science from the Karlsruhe Institute of Technology (KIT) and is currently doing a doctorate in the area of software quality analysis regarding the maintainability at the KIT. During his study, he worked for a leading multimedia company and some startups.

Sebastian Abeck is Professor at the Karlsruhe Institute of Technology (KIT). He is head of the research group Cooperation & Management (C&M) which is doing research and teaching in the area of service-oriented and mobile web applications. His current research topics are: systematical requirements analysis, quality metrics, accessibility, security patterns for identity and access management.

Chapter 4
Knowledge Elicitation and Conceptual Modeling to Foster Security and Trust in SOA System Evolution

John W. Coffey, Arthur Baskin and Dallas Snider

Abstract Software systems based upon Service-Oriented Architecture (SOA) are often large, heterogeneous and difficult to understand. Evolving such systems presents some unique challenges. For example, it is critical to understand the impacts on trust relationships and security as SOA systems evolve. A substantial body of work exists on the idea of knowledge elicitation and management through the creation of knowledge models, which are created to represent the conceptual knowledge of experts. Knowledge modeling based upon concept maps is an efficient process and knowledge representation scheme that holds potential to assist planning in evolving SOA systems. This chapter contains two examples of knowledge modeling in support of SOA system evolution. The first example is an academic study that illustrates the use of knowledge modeling to create a software security assurance case. The second example, which is the main focus of this chapter, pertains to the ongoing evolution of a large, real-world Sustainment Management System software suite named PAVER™. This software is being modified to allow third-party add-in functionality to interact with the base system and to create a SOA federation with other enterprise systems. This article contains a description of a knowledge elicitation and modeling effort to identify trust concerns as this increasingly large and complex federation evolves.

4.1 Introduction

Service Oriented Architecture (SOA) technologies provide new possibilities for the creation of more flexible and interoperable heterogeneous systems. However, with these enhanced capabilities come new and complex issues pertaining to system

J.W. Coffey (✉) · D. Snider
Department of Computer Science, The University of West Florida,
Pensacola, FL 32514, USA
e-mail: jcoffey@uwf.edu

A. Baskin
Intelligent Information Technologies, Indianapolis, IN 46216, USA

© Springer International Publishing Switzerland 2016
E. El-Sheikh et al. (eds.), *Emerging Trends in the Evolution of Service-Oriented and Enterprise Architectures*, Intelligent Systems Reference Library 111,
DOI 10.1007/978-3-319-40564-3_4

41

security and to the issue of whether or not a service one might employ is trust-worthy. A major goal of SOA is to foster the creation of more agile and flexible systems which implies that system evolution might occur at a rapid pace. If a system evolves strictly through software extensions to create new functionality, it will likely evolve more slowly than if a system evolves through the recombination of (potentially newly discovered) pre-defined services. Thus, having a clear understanding of security and trust issues is critical in rapidly evolving SOA systems. Concept mapping [1] and knowledge modeling [2] hold potential to provide efficient ways to capture and represent knowledge and reasoning about security and trust issues in SOA systems. Some preliminary work [3, 4] has demonstrated the utility of concept mapping and knowledge modeling in SOA system development.

Concept mapping provides a visual representation of structured knowledge that allows people to organize and express knowledge of a domain such as security or trust. Subject matter experts (SMEs) can collaborate with knowledge engineers (KEs) in a knowledge elicitation process to create concept maps in order to address a security or trust issue. A substantial body of work exists on the creation of knowledge models, which are created to represent the conceptual knowledge of experts. For some examples, see [5–8]. Knowledge modeling based upon concept maps is an efficient process and knowledge representation scheme.

This chapter contains two examples of concept mapping to address security concerns and trust issues in SOA systems. The first example is an academic study that illustrates the use of knowledge modeling to create a security assurance case [3], a means of demonstrating that security concerns have been met for a software system. This example illustrates the idea of collaborative knowledge modeling with a security expert, the software developer and a knowledge elicitor in order to create the security assurance case. The second example is the main focus of this chapter. A significant ongoing evolution of a large, real-world Sustainment Management System software suite named PAVER™ is to allow third-party add-in functionality to interact with the base system and to create a SOA federation of PAVER™ with other enterprise systems. A knowledge elicitation and modeling effort was carried out in order to address these trust concerns.

The remainder of this article includes a review of literature on security and trust models in SOA systems. A review of the use of concept maps and knowledge models for the capture and representation of structural knowledge follows. Strategies for the elicitation and representation of structured knowledge are discussed. Following this groundwork are the two case studies. The article concludes with a discussion of the work presented and some conclusions.

4.2 Security and Trust in SOA Federations

Security and trust are two related concerns in SOA system evolution. Rasmussen and Jansson [9] stated that security and trust concerns must be treated together to provide protection for Internet commerce. They consider traditional security measures such as authentication and access controls as hard security while using the term soft security to describe the concept of trust and the use of reputation systems that allow for the recognition and approval of services to be utilized. Thomas et al. [10] motivated the interrelatedness but dissimilarity of these two ideas by describing security concerns across trust domains. They cite the commonly occurring scenario in federated SOA systems of the authentication of a service user within the user's local trust domain, and the need for the service provider to trust the authentication performed by the user's identity provider. They described a probabilistic trust level model that seeks to quantify the level of trust that can be placed in an authentication event.

Many of the trust models for SOA services cite the work of Marsh [11] who conceptualized trust as something that could be computed and modeled. In his seminal research, Marsh created formulae to estimate trust between agents regardless of whether the agents are humans or machines. He also formalized levels of risk, competence and cooperation. He described the need to consider how trust levels change over time. He envisioned that his work could be applied to sociology, psychology and in a distributed artificial intelligence environment. Marsh realized that his publication was a work in progress and welcomed other researchers to extend his formulae.

Skopik et al. [12] extended the work of Marsh by using previous experiences and crowdsourcing techniques to create a rule-based approach to establish trust among humans and services. Their approach utilized both qualitative and quantitative measures. Some of the sources for their model include interactions, profiles, structural relations and hierarchies, and manually declared relationships. They utilized directed graphs where the vertices of the graph are the actors and the edges represent the trust relations among the actors. Because there can exist both human and non-human actors in a dynamic SOA system, Skopik et al. established the following five classes of measures to assist in building flexible rules that can adjust with the environment: interaction among the actors, similarity of attributes among the actors, trust relations (e.g., personal, bidirectional, and temporal dynamics), collaboration (e.g., reciprocity and response times), and group statistics such as mean and distribution of directed graph metrics.

Cayirci [13] presented a mathematical model which calculates both trust and risk for services, that provides modeling and simulation as a service (MSaaS). The trust and risk model takes into consideration not only security but also the quality of service and grade of service as stated in service level agreements. The goal of Cayirci's research was to build a model that can help determine if a service has enough credibility and accountability, thus offering protection for both the provider and the consumer of the service.

Kovac and Trcek [14] examined trust in SOA services from a socio-cognitive perspective and presented a model using the following input attributes: temporal dynamics between agents, dependencies on feedback from other agents, the context of the trust relationship between agents, and knowledge gained from prior experiences. The components of their trust model include directed graphs representing social networks, a set of trust values and operators, an algorithm to compute trust and a set of policies that define trust interactions. Their architecture incorporates a global trust web service, distributed trust engines and a web-based user interface. Security in SOA systems requires authentication and fine-grained level of access control. Nair et al. [15, 16] proposed a federated web services framework that would utilize hardware virtualization to provide detailed management of group level permissions, data security and process isolation.

From the previous discussion, one might conclude that both security and trust must be treated as relative terms (relatively trustworthy, poorly secured, etc.) as opposed to two-place predicates (trusted or not, secure or not). Furthermore, the level of security and trust in a federated system can change significantly over time. One might view security as starting from a hardware/software perspective and trust as originating as a human concern to be encoded, somehow, in software. Both present significant ongoing challenges for SOA system developers.

4.3   Concept Maps, Knowledge Models, and Knowledge Modeling

Concept maps [1] are comprised of concepts and linking phrases that make explicit the relationship between concepts. A *concept* is defined by Novak as a perceived regularity in events or objects, or a record of events or objects, designated by a label. Triples comprised of [concept—linking phrase—concept] form propositions which are described as fundamental units of knowledge. For instance, the proposition "knowledge generation presupposes and extends structural or conceptual knowledge" is comprised of the concepts "knowledge generation" and "structural or conceptual knowledge" and the linking phrase "presupposes and extends." Concept maps make manifest the principles of subsumption and progressive differentiation from Ausubel's [17, 18] Assimilation Theory in which more specific concepts are differentiated from and subsumed under more general ones.

Early work on concept mapping for knowledge elicitation was conducted at the Institute for Human and Machine Cognition [19]. Since then, several groups have embraced concept mapping for knowledge elicitation. McNeese et al. [20, 21] used concept maps to externalize the expertise of pilots regarding their decision-making strategies, to create conceptual designs of cockpits, and for internal information management systems and procedures. Knowledge elicitation with concept maps has been used as a means of idea generation in groups of people [22, 23]. Beyond the use of concept maps as a learning and teaching tool, successful applications of concept mapping include institutional memory preservation [6, 24], sharing of domain expertise for public outreach [5], visualization of artifacts in Service-Oriented Architecture (SOA) composite applications [25], representation of software security assurance cases [26], and many other uses.

Building concept maps in electronic format with a tool such as CmapTools [27] offers capabilities to connect the maps via navigational links and to associate supplementary resources such as video clips, images, and links to Web sites with the concepts in the maps. A hierarchical, ordered collection of interconnected concept maps and their accompanying resources pertaining to a given topic comprise what is called a knowledge model [2, 5, 6, 24]. Moon et al. [8] have created a comprehensive resource pertaining to applications of concept mapping and knowledge modeling.

4.4 Studies in Knowledge Modeling for SOA Security and Trust

This section contains two examples of the use of knowledge elicitation and knowledge modeling for SOA systems. The first study illustrates a process of developing a security assurance case for a newly-developed composite application. The second study illustrates the use of knowledge modeling to address security and trust concerns pertaining to a large sustainment management system (an enterprise-level system used to track a large collection of physical assets and to foster decision-making regarding how to maintain them) as the system evolves with the addition of third-party add-in functionality and is more extensively integrated into a federation with other enterprise systems.

4.4.1 Developing a Security Assurance Case Through Knowledge Modeling

Agudo et al. [28] stated that security assurance cases are meant to provide evidence that a system meets specific criteria in order to create confidence in the system's security. Techniques for building such cases range from informal to semi-formal to highly formal. Agudo et al. advocated for semi-formal approaches that encompass both functional and non-functional requirements. The current study would properly be characterized as a semi-formal approach focused on both functional and non-functional requirements. The work was performed by teams comprised of faculty and Master's level software engineering students at a regional university. The work pertains to the development of a SOA system pertaining to a flight reservation system.

4.4.1.1 Motivation

Developing comprehensive security cases is prohibitively expensive in all but the most security-critical software. The goal of the work in this case study was to assess the viability of the idea of developing a set of the most critical security concerns to be addressed using a knowledge modeling approach and then developing the case with three key participants: a security expert, the software developer, and a knowledge engineer. The current approach represents an attempt to bring efficiencies to the security assurance development process, making it feasible to develop a case considering only the major concerns of the security expert.

4.4.1.2   Methods and Results

Planning for the development of the security assurance case proceeded in parallel with the actual software development. The process of developing the security assurance case using this approach is conceived as being comprised of two phases:

Phase 1:   The security expert and knowledge engineer collaborate to create a concept map identifying issues to be addressed in the security assurance case.

Phase 2:   The security expert, knowledge engineer and software developer create a concept map that constitutes the actual security assurance case.

Planning for Phase 1 of the process is carried out by building a concept map pertaining to the particular security concerns of interest to the security expert. In the current study, informal discussions were carried out prior to the building of the concept map which was performed solely by the knowledge engineer. In the general situation, interactive development of the concept map in collaborative work between the knowledge elicitor and the expert would supplant the preliminary informal discussions. Figure 4.1 contains the concept map that was developed to reflect the results of the planning process.

The box in the top-left corner of Fig. 4.1 reads *"How are the design and implementation vulnerabilities being addressed in a Security Assurance case?"* This item is called a focus question—a question that is posed at the beginning of the collaborative mapping process in order to ensure that the knowledge elicited remains relevant to the issue being explored. The goal of the knowledge elicitation effort is to answer the focus question. Making the focus question explicit by recording it in the concept map helps collaborators stay on topic and avoid irrelevancies in the elicited knowledge.

In the concept map in Fig. 4.1, one can see that the software security assurance case will involve both a design touchpoint and an implementation touchpoint—the two times in the development cycle when security issues would be examined under the chosen protocol. This concept map also reveals that the rationale for the case rests with best practices and involves four specific vulnerabilities (in bold near the center of the map) to be addressed:

- Error handling vulnerabilities,
- The need for transaction logs,
- Ensuring that no embedded authentication data is present,
- Allowing minimal privileges for proper operations.

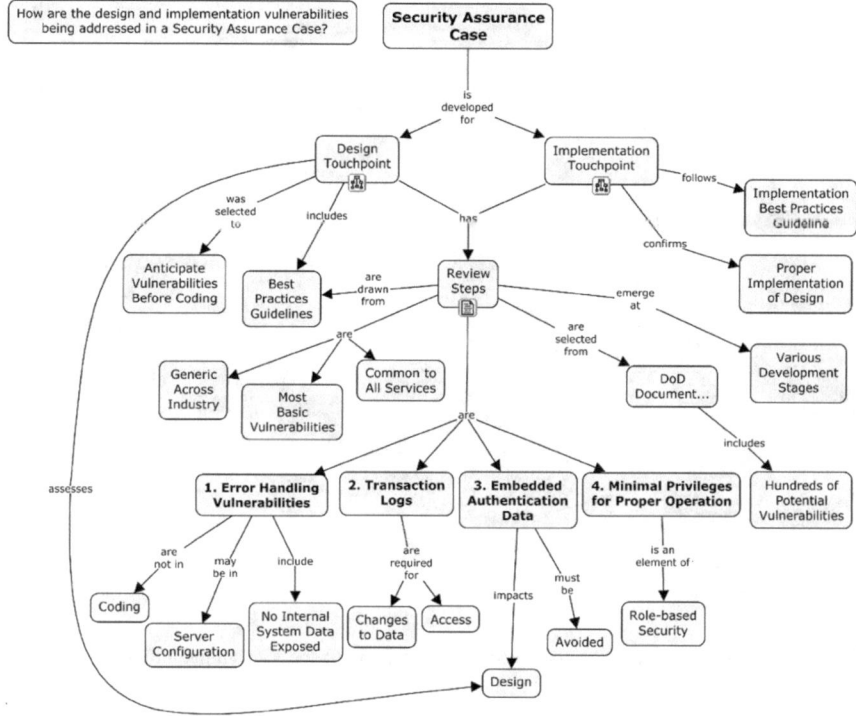

Fig. 4.1 A concept map showing the plan for the security assurance case

In Phase 2, two knowledge engineering sessions were conducted with the software developers, one after the product's design phase, and the other after the product's implementation phase. Figure 4.2 contains one of the concept maps that was developed as part of the security assurance review of the design. At the top of Fig. 4.2, one can see the specific vulnerabilities that appear repeated from the phase 1 planning concept map. Below them is specific documentation regarding how these security issues were addressed.

The review and knowledge modeling effort comprised only a small increment in the overall time required to design and implement the software. The main additional cost of employing this approach was the time of the three participants to conduct the interviews. Each interview lasted approximately an hour. If the total work in design and implementation of the project were 40 h (a very conservative estimate), the security assurance review sessions only added 5 % to the total time required. The resulting concept map-based security assurance case was judged to be a very good communication tool for reviews and other subsequent uses by both the students in the class and by some industry experts who evaluated the work.

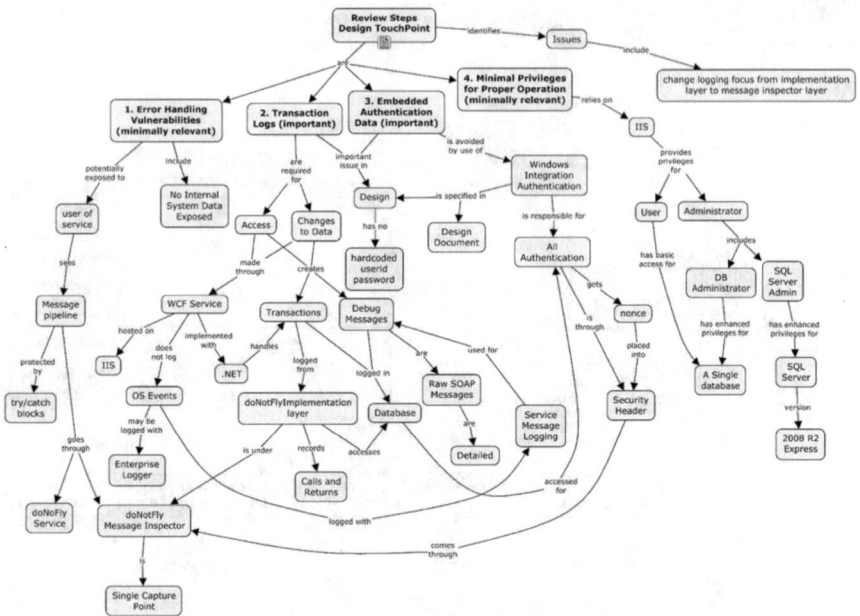

Fig. 4.2 The software security assurance case pertaining to the four vulnerabilities

4.4.2 Assessing Trust Needs for a SOA Federation

4.4.2.1 Motivation

A large-scale sustainment management system named PAVER™ is used by the Department of Defense and NATO to track pavement and roadway inventories and conditions in airports. PAVER™ is being integrated with third-party "add-ins" which are other applications that enhance the functionality of the base product. Additionally, the system is being integrated more extensively with other enterprise systems including a GIS system and other sustainment management systems such as systems that track inventories and condition of railroad track and building roofs. The increased integration of these systems creates federations of interoperating SOA systems at a variety of levels. The goal of the work reported here is to identify trust issues across these levels and to make progress in identifying ways that these issues can be resolved.

The concept of federation in this context can be considered at three different levels of abstraction (from highest to lowest):

- Enterprise Federation—the highest level; PAVER™ and a second system named PCASE are federated with other enterprise systems
- Desktop Federation—an integrated sustainment management system hosting PAVER™, PCASE and add-ins
- Tool Federation—PAVER™ component services interact with add-ins.

4.4.2.2 Methods

Two rounds of knowledge elicitation were conducted in this study. The first round focused on the Desktop and Tool Federation levels of abstraction. The second round focused on the enterprise level of abstraction. Two participants were involved in the work, a SME and a KE. Both the KE and the SME were familiar with this approach to knowledge modeling since they had collaborated on a previous project.

A preliminary informal discussion spanning approximately one hour was devoted to the exchange of background information that was pertinent to the problem areas to be addressed. The knowledge elicitation efforts were quite efficient in terms of human resources—only a single knowledge elicitor working with a single expert. Since the participants were not co-located (one was in Florida and the other in Indiana), WebEx was used with a shared desktop so that both participants could view the emerging concept maps.

The groundwork laid by the expert and knowledge engineer yielded two initial focus questions for round one:

1. What capabilities might we develop to increase trust between PAVER™/PCASE and the add-ins?
2. What are the major concerns that need to be addressed in the management of trust relationships among the add-in vendors, and how do we address them?

A second knowledge elicitation round had a focus on the Enterprise level of abstraction. This level pertains to the entire PAVER™/PCASE + add-ins composite architecture interoperating with other enterprise systems. For instance, PAVER™/PCASE can be used in conjunction with a Global Information System that tracks locations of government property. The second round started with a broad focus question regarding trust issues at the enterprise level. After eliciting a concept map pertaining to the enterprise level of federation, several additional focus questions were identified and an additional concept map was created.

Each session yielded one concept map. After each knowledge elicitation session, the knowledge engineer carried out a refinement process of cleaning up typographic errors, refining or clarifying the linking phrases, generally "wordsmithing" the concepts and linking phrases, and rearranging the structure of the concept maps to enhance clarity and simplicity of reading. After cleaning up the maps, the knowledge engineer passed the maps to the expert who evaluated them for completeness and correctness.

4.4.2.3 Results

In the first concept mapping session, spanning two hours, two concept maps were elicited. In the second round, two more maps were created. Table 4.1 contains a few statistics on the concept maps that were elicited in the sessions.

Table 4.1 The inventory of concept maps that were elicited in the sessions

Map	Concepts	Linking phrases
Round 1		
Capabilities	47	51
Concerns	39	44
Round 2		
Enterprise trust issues	49	53
Authoritative source	42	49

The first knowledge elicitation round enabled the identification of additional focus questions that allow "tunneling in" on issues at a greater depth. While none of them was explored within the scope of the current project, the following focus questions were identified:

1. What are alternative means of resolving trust issues surrounding the numerous file formats at the various layers of federation?
2. How do we resolve competing industry and Department of Defense (DoD) requirements and the corresponding trust issues (in general, or for the file format issue specifically)?
3. In light of trust issues, how should we approach choosing add-ins for direct access to each other through the desktop?
4. How should we assess and foster trust on the file reader issue?

From the first session in the second round on trust at the enterprise level, two additional focus questions were identified:

1. What should be done in the short run and in the long run when mandated authoritative source differs from the operational data source?
2. How can we foster trust in summary data within the enterprise federation?

Construction of the map in Fig. 4.3 yielded several key ideas. First, that trust is fostered through a family of methods supporting discovery, orchestration and composition in SOA applications. After that, the key components in the system were identified. As can be seen in Fig. 4.3, a key idea is that, for the management of trust among add-ins, two key elements are needed: a host object and a window into a database.

Another concept that was captured was the idea of host objects as those responsible for the registration of callback routines for the add-ins. Additionally, the concept of the database adapter as a helper object to the host object was captured. The issue that host objects need to know if the implementation is local in a Windows machine or Web-based (but the database helper does not) was made explicit. Another key idea was that interface contracts foster trust.

An important design principle which the SME characterized as internalization versus externalization was identified. The issue is the degree to which the host object must know about the inner workings of add-ins. If functionality is internalized, the add-in has all the information it needs to perform its responsibilities.

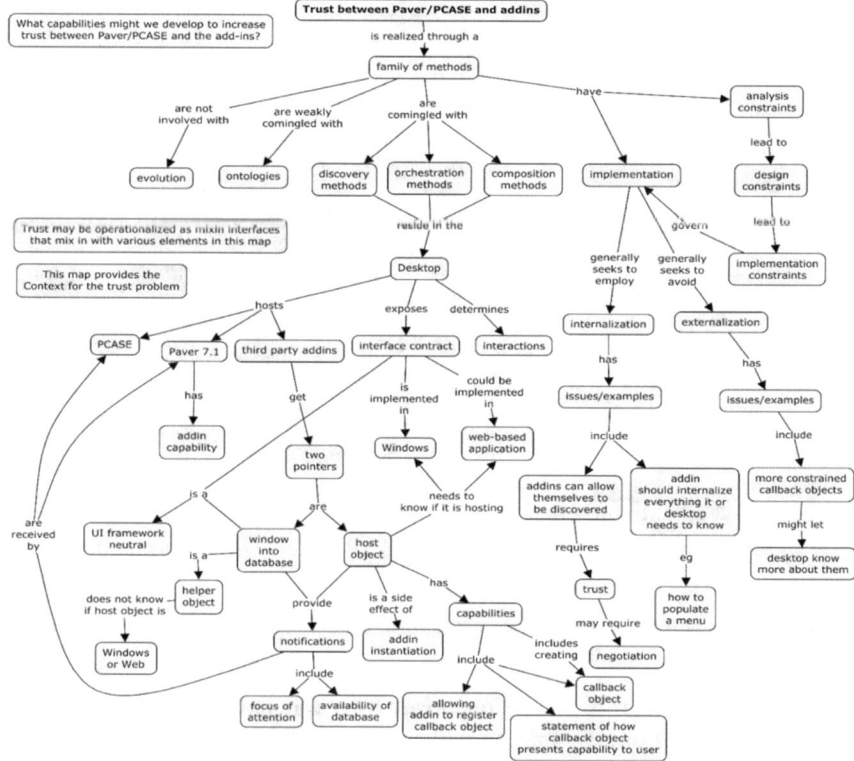

Fig. 4.3 A concept map pertaining to fostering trust between PAVER™/PCASE and the add-ins

The alternative, externalization, requires the host object to be aware of internal functioning of the add-in which yields a higher degree of coupling between the host and the add-in. The more desirable avenue is to limit how much the host object must know about the inner workings of the add-in (to limit externalization) and to have the add-in internalize as much functionality as possible.

Such an approach yields lower coupling from a software engineering standpoint and fosters trust on the part of the add-in creator. Externalization fosters greater coupling, and greater dependency on the host object to "keep doing the right thing." This approach is bad from the perspective of the add-in developer. The reader should refer to Fig. 4.3 to appreciate how these and other ideas are represented in elicited concept map form.

Figure 4.4 contains the second concept map elicited in the first round. As can be seen in Fig. 4.4, this session successfully identified several issues pertaining to the integration of add-ins including how to make an add-in available—to install or bring it into scope through the desktop. The trust relationship is established by the add-in's presence in the desktop since administrative privileges are now necessary

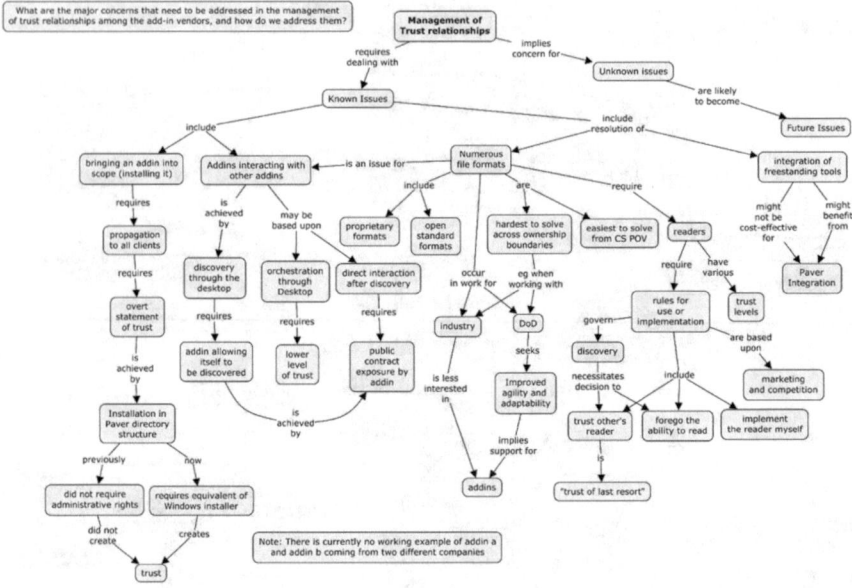

Fig. 4.4 A concept map on the management of trust relationships at the tool and desktop levels

to its installation there. A second important issue was how to have trusting relationships as add-ins directly interact without PAVER™ as an intermediary.

The numerous file formats for data complicate trust relationships at these levels, and the issue of identifying trusted readers that correctly read data in any of the various formats in which data might be stored is a significant concern. The divergent positions of industry partners and the DoD on interoperability with add-ins was identified in this round. The DoD favors increased agility and adaptability in the system by the enhanced use of add-ins whereas industry partners are more conservative on trust issues. For instance, one partner asserted that the only sufficient basis for trust would be through acquisition and direct control of PAVER™.

A second round of knowledge elicitation sought to address trust issues at the enterprise federation level. This level involves interoperation of other enterprise systems with the PAVER™/PCASE + add-in system. Figure 4.5 contains the first concept map that was created. As can be seen in Fig. 4.5, a significant trust issue at the enterprise level pertains to the authoritative source of data when multiple systems contain their own data stores for the same data. In actual system operation, there is a mandated authoritative source and an operational authoritative source, which ideally should be the same. However, in actual systems, the mandated source might be different than the actual or operational authoritative source if the operational source has better data than the mandated source.

Two examples of this disparity were identified. The tagging of pavement by facility ID and GIS data were both identified as instances in which the mandated

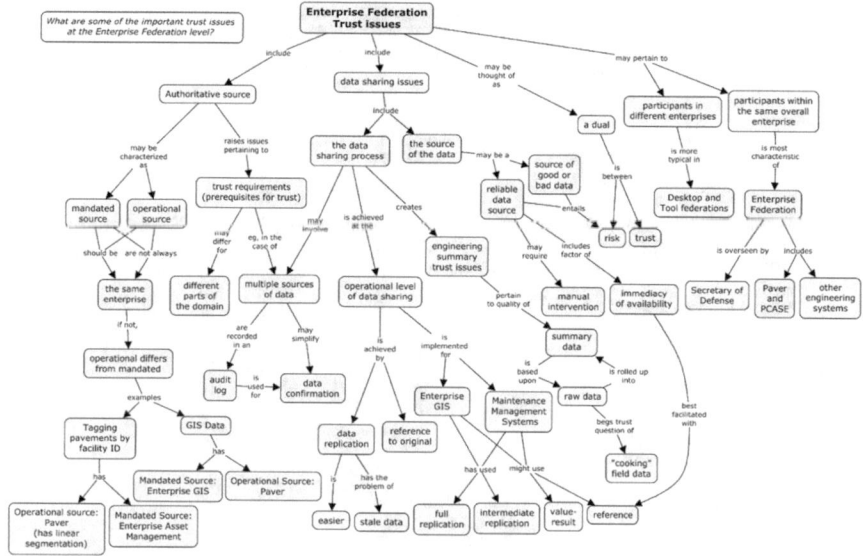

Fig. 4.5   A general concept map on trust issues at the enterprise federation level

and operational authoritative source are different. A subsequent round of KE was conducted to focus on the range of options for dealing with such a disparity.

Data replication versus reference back to a single original data source was another significant concern that was identified. Historically, different levels of replication (full, partial, or none) have been employed. Elicitation of heuristics for when to employ each of these approaches forms a suitable focus question for subsequent analysis. A third major data concern was fostering trust in summary data. Different systems have varying quality of data and employ different algorithms to summarize their data. The quality of these summary data and how these summaries are generated creates another trust issue.

Figure 4.6 is a concept map about the trust issues that arise when a mandated authoritative source of data is different from the operational or actual authoritative source. A root cause of this problem is the historical lack of a clear incentive for the mandated source to be vigilant in incrementally updating the data. A field engineer is usually the one who detects the problem when the data from the software is at odds with conditions at a physical facility s/he is visiting. A trust issue arises when the field engineer detects such a disparity.

A second trust issue pertains to the field engineer getting properly updated data back to the mandated authoritative source. The disparity between mandated and operational authoritative source has been an ongoing problem which is finally improving due to the imposition of sanctions on the mandated source when it does not maintain good data. The KE session led to the identification of trust

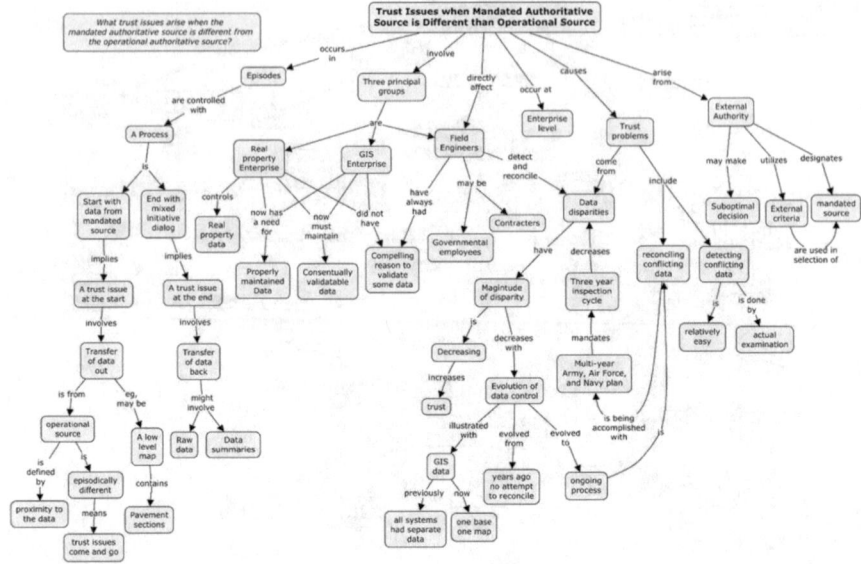

Fig. 4.6 A concept map on trust issues when mandated and operational authoritative source are not the same

relationships between the DoD and PAVER™ and between PAVER™ and the other industry enterprise systems. Efforts made to foster and maintain those relationships are ongoing and sensitive in nature, and therefore will not be discussed in depth in the current work.

The knowledge elicitation work resulted in the creation of five concept maps, four of which are presented here. The relationships among the ones presented in this work are shown in Fig. 4.7 which contains a graphic of a "map of maps." In the actual knowledge model, the icons on four of the nodes are actually dropdown menus that afford access to the concept map on that topic. The two concept maps pertaining to the Tool and Desktop federation level were both at the same level of generality. The second concept map in the Enterprise level (Authoritative Source) illustrates an effort to "tunnel in" on an issue of importance from the more general concept map "Enterprise Federation Trust Issues." Accordingly, that concept map is included at a lower level in the hierarchy.

The purpose of the current work is to illustrate how knowledge elicitation and modeling can lead to a greater understanding of issues in SOA evolution through a systematic examination of the known issues, not to carry out a comprehensive analysis of all issues pertaining to the evolution of this system. In a more comprehensive effort, a similar tunneling in would be performed on a number of additional trust issues.

Fig. 4.7 A map of maps for the concept maps created in this knowledge elicitation effort

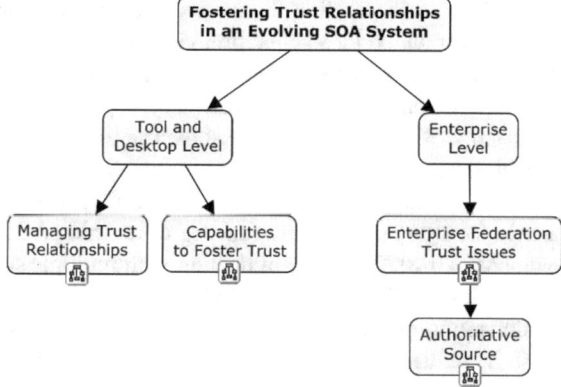

4.5 Discussion

The two studies in knowledge elicitation presented here, for security and trust, addressed issues at two levels. The first example demonstrated specific and low-level documentation of security issues and how they were addressed in developed software. The second example illustrates how high-level abstract conceptual knowledge of broad strategies and approaches to resolve trust issues can be elicited. The two levels of focus are not disjoint, and either can feed into the other.

For example, the security criterion "allowing minimal access privileges" from the software security assurance case, requires a specific implementation that affords different levels of access (e.g.: none, read only, read/update) to various parts of the system and for various categories of users. Actually deploying the system and making decisions regarding who has a given level of access to what requires that trust decisions be made. Another way of phrasing the question of minimal privilege is "What are the relevant trust domains for this system, what is the level of trust we assign to users in each domain, and how much access should each user level have?" Operationalizing a question in such a manner allows use of expert knowledge elicitation techniques to support decision-making. Knowledge modeling for security at a software implementation level can give rise to the need to answer higher-level human-centric conceptual questions such as those about trust, and vice versa.

The work described here evidences progress on capturing known security and trust issues across both case studies. However, it might be concluded that little success occurred in anticipating currently unknown issues, either in the software security assurance case or in the trust work. It is well known that anticipating an unknown security threat is a significant problem and that zero-day attack vulnerabilities [29] are usually only understood after-the-fact. With a relatively stable and definable set of current and potential trust domains, these domains can be identified through knowledge elicitation, and trust relationships can be constructed via either ad hoc methods, or by knowledge elicitation. However, in a rapidly evolving federation, understanding the evolving trust relationships is an inherently difficult problem for which idea generation via knowledge modeling is one potential answer. No claim is made that knowledge modeling is a panacea for this problem, but it is at least one potential partial answer.

Results of this work provide some confirmation for the claims pertaining to the fractal nature of SOA applications across different levels of abstraction presented in Baskin, Reinke, and Coffey (in this volume). For example, the new work in this article focuses on trust, and (among other things) authoritative source as a critical element of trust. The issue of dealing with multiple sources of the same data and the need to know which is the one to be most trusted, and trust issues pertaining to data reader modules cross-cut all levels of abstraction. At the tool and desktop levels trust in readers and core tools maintaining interoperability with add-ins was a critical concern. Authoritative source and how to access it was the issue. At the enterprise level, the issue of the mandated authoritative source being different than the operational authoritative source was an important concern. While this paper has a narrower focus than the companion paper, similar validations could be obtained with methods advocated here.

4.6   Conclusions

The first study demonstrates several advantages that arise from use of the proposed scheme for the development of security assurance cases. The knowledge engineering approach facilitates the introduction of assurance case touch points into the development process in a way that does not entail a burdensome increment in work load. Participation in the interviews was shown to improve programmer sensitivity to potential software vulnerabilities. The concept map security assurance cases would be routinely linked to other documentation such as design rationale, to provide a view of the software structure in which the security concerns are seamlessly integrated. Finally, the visual nature of concept map-based security assurance cases could facilitate communication with diverse project stakeholders. While the security assurance case described here pertained to initial software development, the same approach with the more narrow focus on security issues pertaining to changes to a system could be made as a system evolves.

The second study yielded a knowledge model that provides a global view of the major trust issues at three levels of abstraction within a large, evolving, real-world SOA federation. The knowledge model provides an historical context regarding how data replication versus single-source data has been utilized, it establishes operating principles for the incorporation of add-ins to the code base of a large sustainment management system and for trust concerns of both industry partners and governmental agencies that are forming an increasingly integrated enterprise-level federation. These results were achieved in a personnel-efficient manner with only two participants, a knowledge engineer and a subject matter expert and in a time-efficient way. While discretion pertaining to the ongoing nature of the evolving trust relationships involved here limits the authors' ability to delve into details, an improved awareness of important issues was gained as a result of the knowledge elicitation sessions.

References

1. Novak, J.D., Gowin, D.B.: Learning how to learn. Cambridge University Press, New York (1984)
2. Coffey, J.W., Cañas, A.J., Reichherzer, T., Hill, G., Suri, N., Carff, R., Mitrovich, T., Eberle, D.: Knowledge modeling and the creation of El-Tech: a performance support and training system for electronic technicians. Expert Syst. Appl. **25**(4), 483–492 (2003)
3. Coffey, J.W., Snider, D., Reichherzer, T., Wilde, N.: Concept mapping for the efficient generation and communication of security assurance cases. In: Proceedings of IMCIC'14, Orlando, FL. 4–7 Mar 2014, pp. 173–177. ISBN-978-1-936338-97-9
4. Coffey, J.W., Baskin, A., Reichherzer, T., Wilde, N.: Recovering SOA system architecture from low-level artifacts with a semi-automated approach involving CARET and knowledge elicitation. Int. J. Softw. Eng. Knowl. Eng. **26**(1) (2016, Jan) (to appear)
5. Briggs, G., Shamma, D., Cañas, A.J., Scargle, J., Novak, J.D.: Concept maps applied to Mars exploration public outreach. In: Cañas, A.J., Novak, J.D., González, F. (eds.) Concept Maps: Theory, Methodology, Technology. Proceedings of the First International Conference on Concept Mapping, pp. 125–133. Pamplona, Spain (2004)
6. Coffey, J.W., Eskridge, T.: Case studies of knowledge modeling for knowledge preservation and sharing in the U.S. nuclear power industry. J. Inf. Knowl. Manage. **7**(3), 173–185 (2008)
7. Coffey, J.W., Hoffman, R.R., Cañas, A.J.: Concept map-based knowledge modeling: perspectives from information and knowledge visualization. Inf. Vis. **5**, 192–201 (2006)
8. Moon, B., Hoffman, R.R., Novak, J., Canas, A. (eds.): Applied Concept Mapping: Capturing, Analyzing, and Organizing Knowledge. CRC Press (2011). ISBN 9781439828601
9. Rasmusson, L., Jansson, S.: Simulated social control for secure internet commerce. In: Proceedings of the 1996 Workshop on New Security Paradigms (NSPW '96), pp. 18–25. Lake Arrowhead, CA (1996)
10. Thomas, I., Menzel, M., Meinel, C.: Using quantified trust levels to describe authentication requirements in federated identity management. In: Proceedings of SWS'08, October 31, 2008, Fairfax, Virginia, USA, pp. 71–79. ACM 978-1-60558-292 (2008)
11. Marsh, S.P.: Formalising trust as a computational concept. Stirling, Scotland: Ph.D. dissertation, Dept. Computing Science and Mathematics, University of Stirling (1994)
12. Skopik, F., Schall, D., Dustdar, S.: Modeling and mining of dynamic trust in complex service-oriented systems. Inf. Syst. **35**, 735–757 (2004)
13. Cayirci, E.: A joint trust and risk model for MSaaS mashups. In: Proceedings of the 2013 Winter Simulation Conference, 8–11 Dec 2013, Washington, D.C, pp. 1347–1358
14. Kovac, D., Trcek, D.: Qualitative trust modeling in SOA. J. Syst. Architect. **55**, 255–263 (2009)
15. Nair, S.K., Djordjevic, I., Crispo, B., Dimitrakos, T.: Secure web service federation management using TPM virtualisation. In: Proceedings of the 2007 Secure Web Services Workshop (SWS'07), pp. 112–121. Fairfax, VA (2007)
16. Nair, S.K., Djordjevic, I., Crispo, B., Dimitrakos, T.: Secure web service federation management using TPM virtualisation. In: Proceedings of the 2007 Secure Web Services Workshop (SWS'07), 2 Nov 2007, pp. 73–82, Fairfax, Virginia, USA
17. Ausubel, D.P.: Educational Psychology: A Cognitive View. Rinehart and Winston, New York (1968)

18. Ausubel, D.P.: The Acquisition Retention of Knowledge: A Cognitive View. Kluwer, Dordrecht (2000)
19. Ford, K.M., Cañas, A.J., Coffey, J.W.: Participatory explanation. In: Proceedings of the Sixth Florida Artificial Intelligence Research Symposium (FLAIRS '93), Ft. Lauderdale, FL, Apr 1993. pp. 111–115
20. McNeese, M., Zaff, B., Brown, C., Citera, M., Selvaraj, J.: Understanding the context of multidisciplinary design: establishing ecological validity in the study of design problem solving. In: Proceedings of the 37th Annual Meeting of the Human Factors Society, 1993. Santa Monica, CA
21. McNeese, M., Zaff, B.S., Citera, M., Brown, C.E., Whitaker, R.: AKADAM: eliciting user knowledge to support participatory ergonomics. Int. J. Ind. Ergon. **15**, 345–363 (1995)
22. Novak, J.D.: Learning, Creating, and Using Knowledge: Concept Maps As Facilitative Tools in Schools and Corporations. Lawrence Erlbaum and Associates (1998). ISBN-13: 978-0805826265
23. Coffey, J.W.: Facilitating idea generation and decision-making with concept maps. J. Inf. Knowl. Manage. **3**(2), 1–14 (2004)
24. Coffey, J.W., Hoffman, R.R.: Knowledge modeling for the preservation of institutional memory. J. Knowl. Manage. **7**(3), 38–49 (2003)
25. Coffey, J.W., Reichherzer, T., Wilde, N., Owsnicki-Klewe, B.: Automated concept-map generation from service-oriented architecture artifacts. In: Proceedings of the 5th International Conference on Concept Mapping. Valetta, Malta, Sept 2012
26. Snider, D., Coffey, J.W., Reichherzer, T., Wilde, N., Terry, C., Vandeville, J., Heinen, A., Pramanik, S.: Using concept maps to introduce software security assurance cases. CrossTalk J. Defense Softw. Eng. **27**(5), 4–9 (2014)
27. Cañas, A.J., Hill, G., Carff, R., Suri, N., Lott, J., Eskridge, T., Gómez, G., Arroyo, M., Carvajal, R.: CmapTools: a knowledge modeling and sharing environment. In: Cañas, A.J., Novak, J.D., González, F. (eds) Concept Maps: Theory, Methodology, Technology. Proceedings of the First International Conference on Concept Mapping, Pamplona, Spain (2004)
28. Agudo, I., Vivas, J.L., López, J.: Security assurance during the software development cycle. In: Proceedings of CompSysTech '09, the International Conference on Computer Systems and Technologies and Workshop for PhD Students in Computing. ACM, June, 2009, pp. II.7-1–II.7-6
29. Bilge, L., Dumitras, T.: Before we knew it an empirical study of zero-day attacks in the real world. In: Proceedings of CCS'12, October 16–18, 2012, Raleigh, North Carolina, USA. 2012, pp. 833–844. ACM 978-1-4503-165

Chapter 5
The Fractal Nature of SOA Federations: A Real World Example

Arthur Baskin, Robert Reinke and John W. Coffey

Abstract Fractal concepts are often said to be recursively self-similar across multiple levels of abstraction. In this paper, we describe our experience with the fractal nature of SOA designs for sustainment management tools as these tools evolve into even more dynamic, federated systems that are orchestrated over the web. This chapter summarizes insights gained from more than twenty years of software development, maintenance, and evolution of a major pavement engineering tool named PAVER™. We consider both theoretical and experiential aspects of SOA federations at three levels of abstraction: (1) a loosely coupled federation of enterprise systems with PAVER™ as one member, (2) a tightly coupled federation of two pavement management tools (PAVER™ and PCASE) where each has a separate domain identity and development team, and (3) an emerging federation of plugin tools, which provide additional pavement engineering functionality and can come from competing civil engineering firms. These plugin tools exist at different levels of abstraction within the level of the main system and are, again, fractal. We organize the presentation of our experiences in this domain by describing how SOA elements including Ontologies, Discovery, Composition, and Orchestration are fractal whether we are looking at algorithms or persistent state. We also define and describe a third orthogonal fractal dimension: Evolution. Although the details of the implementation solutions at the differing levels of abstraction can be substantially different, we will show that the underlying principles are strikingly similar in what problems they need to solve and how they generally go about solving them.

A. Baskin (✉) · R. Reinke
Intelligent Information Technologies Corporation, Indianapolis, IN 46216, USA
e-mail: abaskin@intelligent-it.com

J.W. Coffey
Department of Computer Science, The University of West Florida,
Pensacola, FL 32514, USA

© Springer International Publishing Switzerland 2016
E. El-Sheikh et al. (eds.), *Emerging Trends in the Evolution of Service-Oriented and Enterprise Architectures*, Intelligent Systems Reference Library 111,
DOI 10.1007/978-3-319-40564-3_5

5.1 Introduction

It is the goal of this article to make the case that fundamental concerns in the development and evolution of SOA systems are self-similar or fractal across different levels of abstraction in SOA systems. These observations about the fractal nature of federations of service oriented systems (SOA) are grounded in more than twenty years of software development and maintenance of a series of condition-based civil engineering maintenance management systems. Foremost among these systems has been the PAVER™ system, which is a pavement management system for airfields and roadways. In some cases, these decision support tools are targeted at elevating the expertise of a normal practitioner to be closer to that of the expert, whose expertise is embodied in the system. In other cases, these systems support a loose federation of human decision makers, who are again engaged in complex decision making. For example, the PAVER™ pavement management system embodies international standards for ways to convert visual observations of the pavement into a standard pavement condition index, which can be used prioritize the use of scarce resources for civil infrastructure repairs.

The author's experiences with many of the principles, which are described here, predate the general emergence of SOA concepts, and, therefore, we interpret some of the historical material looking backward. To enable this perspective, we provide a brief historical account of PAVER™. We justify this retrospective interpretation of history because we were responding to the needs for modular evolution of separate tools and the need for loose coupling among tools to allow human decision makers to compose data from varying sources to make engineering judgements. We assert that these are some of the same requirements that SOA systems are proposed to address. Increasingly over the last ten years, we have been actively injecting insights from the SOA approach into the structure and organization of our software systems. In fact, the preparation of the material for this chapter has had a direct impact on the emerging support for plugins to the pavement engineering desktop because we have used SOA principles to guide the definition and implementation of the plugin framework, which has overlapped the development of the material for this chapter in time.

In what follows, we try to present both the experiences that drove us to arrive at or validate principles as well as a theoretical basis for the principles where we believe we have found one. Two of the authors (Baskin and Reinke) have been involved with software development tasks where attention to the evolution of the software was part of the problem and the time to reflect on why some development techniques work better than others was available. In some situations, the principles emerged from the experience supporting the evolution of the software and in other situations, we were able to bring principles from other disciplines to bear on our software develop projects.

We have attempted to organize what we have learned at two distinct levels of abstraction: (1) overall principles and (2) origin of the principles. Where possible, we have attempted to provide both a theoretical derivation of the principles and concrete examples of the principles, which are drawn from our experience with pavement management systems. Readers may wish to pursue either the theoretical or historical description of the principles or simply review the summaries of what we believe we have learned.

We have found SOA Federations to be effective for managing complex software systems: Across time with the interplay of (relatively) independent actors, domains of problem solving, and goals (with or without a common owner). Unpacking this deceptively simple statement is our goal in this chapter. Because many of the terms we use do not have precise definitions (or even agreed definitions), we will spend some time in each section describing what we mean by our terms. These definitions will be especially important because we need to use them across several fractal levels of abstraction. Although we focus on the domain of pavement engineering maintenance management systems in this chapter, we have applied these principles in the development of decision support systems in other areas of civil engineering and in shortening time to market for new products by using agile software development techniques to begin to develop software to test a new product even before the design of the new product has been finalized.

Figure 5.1 shows the three fundamental dimensions of our separation of issues for SOA Federations.

In the Fig. 5.1, we identify traditional SOA issues along one axis. At the risk of oversimplification, we can say that ontologies precede and are necessary for supporting Discovery, Orchestration, and Composition. In our analysis, we look at ontologies for procedural and declarative aspects of a problem and the need for evolution of ontologies, which might be the hardest problem of all to solve. The remaining three SOA issues: Discovery, Orchestration, and Composition can be thought of as describing the state of a particular SOA system at one point in time. That is, a working SOA system must discover procedural and declarative elements so that they can be composed and finally orchestrated. In our SOA desktop for pavement engineering, the desktop follows a discovery process each time it is started to determine the components, which are available for composition in this user session. The human user plays a central role in orchestrating the interoperation of some of the components and simplifies the software orchestration problem for us.

Fig. 5.1 Three dimensions of issues for fractal analysis. We separate SOA issues into procedural and declarative approaches to software construction. Finally, we explore the evolution of all of these aspects over time

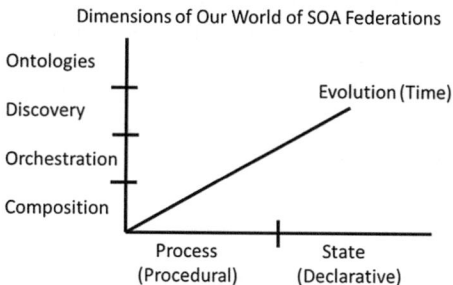

5.2   The Historical Context of This Work

Our work on PAVER™ started with a series of workshops, which were sponsored by the Federal Highway Administration and supervised by Shahin [1] who continues to guide these efforts for the U.S. Army Corps of Engineers, which owns the software. The workshops brought together researchers in pavement engineering, practicing pavement engineers, and our software development team. These workshops guided the development of the software to help define the best practices in the pavement management domain and to embody them in a revised version of the PAVER™ system. These user group meetings continued for years after the startup phase ended and the interplay of the expert users and the software designers has supported the co-evolution of best practices for pavement management and the design of new PAVER™ features [2]. New features have also been driven by pavement engineering research carried out by Dr. Shahin and others. This emergent requirement for software evolution as a consequence of best practice evolution, which was, in turn, partially stimulated by the PAVER™ software system, gives rise to our attention to evolution as a basic area of interest.

The PCASE system evolved largely independent of PAVER™ but was used by many of the same pavement engineers at the same locations. About fifteen years ago, the U.S. Air Force funded an integration program to bring PAVER™ and PCASE together in what the user would experience as a seamlessly integrated system. In addition, each separate system had to be able to be deployed independently as it had been done in the past. A user needed to be able to install either of these civil engineering tools first and then optionally add the other. Finally, this federation of loosely coupled tools had to be developed by two different research groups with two largely disjoint user groups and rates of evolution. Our experience with such a federation of largely independent software development efforts that produce a seamless integration of the tools in the user experience has led to our attention to the issues of discovery, orchestration, and, again, evolution of software systems in federations of civil engineering tools.

In the latest iteration of work on these pavement engineering tools, we have been tasked by the Corps of Engineers to expand the existing tools to provide support for international users, interoperation with other Department of Defense enterprise systems, and to support the incremental evolution of these pavement engineering tools through an enhanced version of the plugin mechanisms, which have been available but underutilized for ten years. We have explicitly used the principles described in this chapter to guide the additions to the plugin machinery and to suggest requirements that might not otherwise have emerged from the domain requirements.

5.3 Literature on SOA Federations, SOA Elements, Algorithms and Data Persistence

In the following section, we provide a brief overview of literature pertaining to SOA federations, the SOA elements we consider to be fractal in nature, and the relationship between the cross-cutting concerns of algorithms and persistent data. An extensive literature is available on the topic of SOA federations. Zdun [3] describes a federation as a controlled environment with a collection of peer services. Each service might be controlled by one or more federations, but within a single federation, peer services can easily interact. Federations are based upon middleware that fosters interoperability of loosely-coupled services. The term federation appears in the literature in two contexts and it has evolved out of two distinct but increasingly interrelated technological bases. The High-Level Architecture, and accompanying Real-Time Interface (HLA-RTI) [3–5] originated in the mid-1990s in the DoD as a platform for real-time simulation. The IEEE 1516 standard grew out of the High-Level Architecture initiative. In this context, a federation is a collection of simulation entities connected and orchestrated by the real-time interface. A federate is an individual IEEE 1516-compliant simulation entity. The HLA includes a federation management API [6].

By the late 1990s, well after early versions of PAVER™ were already in use, SOA was an emergent technology for the creation of composite applications. SOA is based upon W3C standards including WSDL, SOAP, and XML Schema [7–9]. The HLA-RTI notion of federation is somewhat distinct from the more general notion of federation as described in SOA literature because it is a particular use in simulations rather than for the creation of composite applications that accomplish a broad range of business or engineering purposes. However in more recent times, calls for integration of HLA and SOA have occurred [5] as it is viewed that lessons learned in one community might benefit the other. Additionally, benefits might be obtained by the interoperation of both standards. For instance, Seo and Zeigler [10] described the DEVS/SOA system to provide web service-based simulations.

A major concern in the creation of federations involving multiple providers is managing identities and access of federates [11]. Several standards-based protocols have been proposed or implemented to create federations [12–14]. Li and Karp claim that the federated identity management approach has proven to be difficult to use and upgrade, and is not scalable. They state that federation based upon identity is the wrong focus, rather the focus should be on access management instead. They illustrate implementing access control policies using SAML certificates [15]. Thomas and Meinel [16] describe their own management system, which also relies on open web service standards, to provide reliable digital identities. Hatakeyama [17] describes what is termed a "federation bridge" to facilitate cross domain identity federation. Anastasi et al. [18] state that service providers offer their services using proprietary management software, interfaces and virtualization technologies which make interoperability more difficult to achieve. They discuss their simulation tool SmartFed which is designed to simulate cloud-based federations.

Since a detailed review of literature pertaining to the SOA elements of discovery, composition, orchestration and ontology would encompass an entire book chapter in itself, we refer the interested reader to the following representative works. Al-Masri and Mahmoud [19] attempt to incorporate client goals into service discovery queries via a means that would rank candidate services in a manner similar to query result rankings implemented in general-purpose web search capabilities. Dsbrowski and Pacyna [20] address inter-domain service discovery and claim that service discovery systems will require a strongly interwoven identity management component. They state that support for service discovery across service domain boundaries must be implemented in identity management systems in order to provide a safe discovery system between services from different business areas.

Tolk et al. [21, 22] have done important work on composition and orchestration within large governmental SOA federations. They describe current Homeland Security systems which must integrate data and processing capabilities from twenty two previously separate agencies. In [21] they describe model-based, top-down orchestration of heterogeneous Homeland Security systems with discovery and composition of IT capabilities included in a system-of-systems bottom up. In [22] they describe a mathematical model for the selection or elimination of candidate services, and for their orchestration and execution. They describe this work as a first step towards self-organizing federation languages. Work by Rathnam and Paredis [23] provides an ontology-based framework to simplify the reuse of federates in a federation object model. While their work pertains to HLA-RTI federations, it is applicable in principle to SOA systems in general.

We have identified as key issues in the current work, the cross-cutting concerns of algorithms and data persistence in SOA. Calvanese et al. [24] discuss this issue stating that one's view of the pre-eminence of data or process is often a function of one's viewpoint, for instance if one is a business process analyst or a data manager. They cite an article by Reichert [25] which makes the claim that "data and processes should be considered as two sides of the same coin." They discuss "data-aware processes" and conclude that the database theory community has developed the defining techniques to deal with data and processes. They cite several important issues including verification issues pertaining to the modeling of what data can be changed by a process. Dobos et al. [26] describe a platform for the management of reusable process components and for the federation of data stores in order to support data persistence, statistical analysis and presentation of the data.

Data persistence is an important aspect of data management. Krizevnik and Juric [27] cite data persistence problems stemming from poor data quality, heterogeneity of data sources and poor system performance in SOA systems. They describe a SOA persistence model relying on master data management (MDM), and data transfer standardization by the use of service data objects (SDOs) [28] in order to build a data services layer in a SOA system. Software companies seek to build in data services layers in their SOA architecture solutions. For instance BEA Systems [29] describe the AquaLogic Data Services Platform (ALDSP) stating that the environment employs a declarative approach to the construction of data services that are based upon XML Schema for data definitions and XQuery as the service composition language.

Takatsuka et al. [30] state that cloud computing and machine-to-machine technologies require "context-aware" services to deal with heterogeneous data from distributed systems. They describe a rule-based framework to create context-aware services where context is taken to mean situational information that can be true or false. Sarelo [31] describes the HERMES framework for ubiquitous communication management using web services with serialized XML, data replication with peers storing full copies, and simultaneous data update of all replicates. The previous literature review barely scratches the surface of available literature on all these aspects of SOA, but it provides the interested reader with a starting point for further exploration.

5.4   Three Levels of Abstraction for SOA Federations

Although the development of the systems of interest in the current work has evolved from support tools for a single practicing field engineer toward the needs of enterprise systems, we may now look at these emerging systems from the top down. Figure 5.2 shows the three major levels of abstraction for the tools which we discuss in this chapter:

Dr. Shahin's work on condition-based maintenance management systems gave rise to a series of similar system. We put forth RAILER, which is a system for maintenance management of railroad track, and ROOFER, which is a system for maintenance for roofing on buildings, as examples of this more extended family of civil engineering systems. Taken together, these systems, which are focused on largely disjoint assets, form an enterprise level system for the application of best

Fig. 5.2 Three levels of abstraction for SOA Federations: (1) Enterprise Federation, which brings together separate line of business systems for exchange of summary data; (2) Desktop Federation, which brings together more tightly coupled systems with separate identities and intersecting user functionality; (3) Tool Federations, which break individual desktop tools into component services as a technique for managing complexity or ease of extension

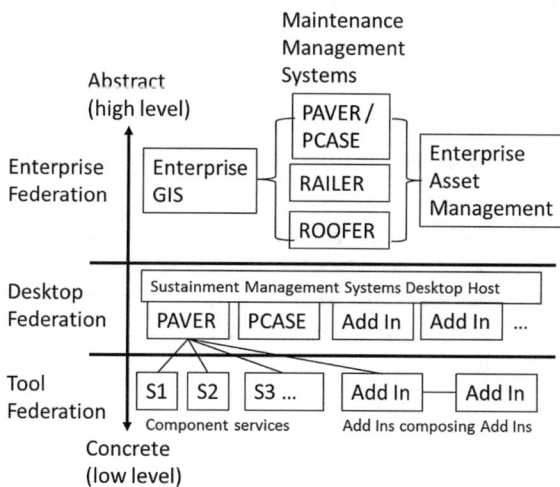

practices for maintenance management so as to provide the safest and most productive use of the civil infrastructure assets for the lowest possible cost. There is increasing demand for interoperability among these separate systems with separately developed enterprise systems for geographical information systems (GIS) and enterprise systems for real property and asset management. GIS systems are used to integrate disparate data bringing together data from separate systems and overlaying these data geospatially. Asset management systems are used to track the value of assets and to plan for allocation and preservation of these assets.

The development of the desktop federation of PAVER™ and PCASE was driven by the user need for these systems to share a common inventory and constrained by the need to allow the tools to retain their separate identities while being able to be seamlessly combined in a single desktop when needed. One of the greatest challenges in this federation was to support the separate rates of evolution of PAVER™ and PCASE and to accommodate the various possible combinations of separate versions.

In its latest incarnation, PAVER™ has been modified to support multiple users and multiple deployment options, which include the traditional standalone install, a thick client-server installation, a shared remote application server, and a web browser interface. As part of managing both the increased pavement engineering capabilities and the varied deployment options, we have used traditional SOA concepts to break the main PAVER™ application into a family of interoperating component services, which can be dynamically loaded in a standalone application or accessed over a web service interface to a remote server. The user is able to switch seamlessly between these modes on each "File/Open" operation, which might open a local database with local services one time and a remote database with remote services the next. This Tool Federation has been extended to support plugins (called Add Ins in PAVER™). Plugins can be a way to extend the core pavement engineering modules and a way to bring together tools from other pavement engineering companies to leverage the common inventory and GIS display tools in much the same what that PCASE can do.

5.5 Dimensions of Our SOA World at Each Level of Abstraction: Real World Example

This section describes our experience with each of the elements in our three dimensional space at each level of abstraction as an extended real world example of these issues at three distinct levels of abstraction. The next section identifies the fractal issues, which have been found to be common to these different levels of abstraction.

5.5.1 Enterprise Federation

The Enterprise Federation of relatively independent systems is a major initiative in the U. S. Department of Defense (DoD) and the work on pavement management systems has been an early emphasis. Unlike the more mature developments at lower levels of abstraction, the Enterprise Federation is still a work in progress; therefore, the results of our analysis at this level are a bit more tentative but we are able to bring lessons learned from the lower levels of abstraction to bear at this level to help guide the software development.

5.5.1.1 Ontologies

Declarative: The PAVER™ system is considered the authority of record for condition-based maintenance of roadway and airfield pavements in the U.S. DoD and for NATO. Using the system, a pavement engineer obtains data about the state of the pavement, its likely future condition, and the cost of foreseeable maintenance. Because the pavement engineer will usually actually stand on or drive over the pavement to make an inventory and condition assessment, the pavement engineer can determine the specific pavement quality and present usage. In addition, the pavement engineer will break the overall pavement extent into manageable sizes for future work planning. The enterprise GIS system and PAVER™ must share a common geospatial rendering of the overall pavement extents (a shared ontology) and PAVER™ can be used to subdivide the overall area into subcomponents, which are the smallest unit of measure on the pavements (called Sections). Sections, again, form a shared ontology between the systems. In this case, the GIS system is the authority of record for the overall pavement extent and PAVER™ is the authority for the section boundaries, which are needed for work planning. In a similar way, the pavement engineer is often in the best place to tag the pavement according to the required asset management attributes, which are called CATCD (category code to summarize type, use, and cost) and RPUID (Facility identifier). Within this rigid standardization, a facility can be divided into "segments" as long as each segment is part of the facility and can be assigned a CATCD. In this second case, the shared ontology is closely controlled by the real property system (CATCD and RPUID) but the segmentation and the assignment of attribute tags to elements of pavement might be shared between the asset management system, which is the authority of record, and the pavement management system because the pavement engineer is more likely to actually stand on the pavement. Definition of these shared ontologies has consumed many hundreds of hours of group meetings, which have involved many more systems than PAVER™, and the implementation of the software support for the emerging ontology has been the simplest part.
　　Procedural: Unlike the shared declarative ontology, which can be shared between the U.S. Air Force, Navy, and Army, the procedural ontology varies by service because the GIS and real property asset systems are not the same. The state

of the procedural ontology for reconciling data at the enterprise level is at the same place that PAVER was more than twenty years ago. There have been many meetings and deliberations collecting required data, reconciling the differences between the mandated authoritative sources for the data and the operational sources for the data. The PAVER system is only one of sixteen major systems participating in this process. The fact that pavement management data must interoperate with different enterprise level systems and be collected according to a variety of business rules means that the ontology must be able to evolve with the emerging changes. As the enterprise data become more available, the best practices as the enterprise level will co-evolve alongside the software systems in the same way that best practices for pavement management and PAVER co-evolved twenty years ago.

Evolution: The enterprise federation brings together different types of structures: vertical (e.g. buildings) and horizontal (e.g. roadways, power grids) for a variety of largely independent entities (e.g. Air Force, Navy, Army, municipalities). The real property/asset management ontology relates physical structures to congressional authorization for funding and is closely controlled. The individual services have evolved separate approaches for the segmentation of assets and the assignment of CATCDs. In the process of integrating sixteen different civil engineering disciplines, of which PAVER™ is just one, the formal ontology for sharing of GIS data has gone through several major versions and version four of the specification is nearing completion and has shown that the services must support evolution of their components driven by their history, current needs, and future missions.

5.5.1.2 Discovery

Declarative: The discovery problem for the enterprise federation is still being solved by ongoing business process reengineering and software development. Each service is developing its own standard operating procedures for how these data should be collected, coordinated, and used for resource allocation and sustainment of the assets. These declarative procedures are tailored to the details of the operations of each service.

Procedural: As a practical matter, these systems are all composed of a union of humans and software systems. In some situations, procedures can be automated and in others, some form of engineering judgement about the interpretation of the data is required. Some agencies have chosen a focus on data replication and some on direct linkage of data. In some situations, the process can be fully automated and in others only semi-automated. The GIS systems and the real property linkage must be able to tolerate the variety of presentations of shared data from each service and/or from each civil engineering domain.

Evolution: The methods for discovery must be able to vary across the services (a form of evolution across situations rather than time) and they must be able to evolve along with changes in technology and best practices. The methods for this type of modular replacement of parts are still being developed, but the problem is clearly present in the variety of approaches to this data alignment problem continues to increase.

5.5.1.3 Composition

Declarative: The PAVER™ inventory has long been "composed" of Networks, which contain Branches, which, in turn, contain Sections. A Section is the smallest unit of maintenance activity and must have both a uniform structure and history. A section may be broken into samples, which allows quicker inspections by extrapolating data from samples to the entire section instead of inspecting everything. The GIS data integration problem is being addressed by the notion of a map which is "composed" of layers. Each layer can be associated with an engineering discipline, e.g. pavements, and data attributes for a region in that layer can come from either geospatial data (e.g. area), asset management (e.g. CATCD), or a maintenance management system (e.g. condition of the asset). A maintenance system is usually expected to rollup lower level data (from segments) into values at the facility level.

Procedural: The procedural aspects of composition at the enterprise level are easiest to see in the GIS presentations, which are data visualization tools at their core. Much of the composition of maps can be interactively driven by a human viewer, who can turn layers on and off as well as selecting among a variety of coloring strategies. In the pavement domain, image capture devices are now able to capture roadway images, which can be analyzed for distresses and dynamically aggregated into "samples," which can then treated like the more traditional samples at the bottom of the pavement composition structure. Rollup of data over the composition structure can be seen as flattening the depth of the composition to get summary data and the flattening algorithms can be weighted by size or importance to mission.

Evolution: The composition of the various engineering maintenance management systems is evolving as some new condition-based maintenance management systems are added and others consolidated. The composition of GIS information in the form of layers and coloring strategies ("themes") is constantly changing as new theme definitions are defined and new attributes added to each layer.

5.5.1.4 Orchestration

Declarative: The prevalence of standard operating procedures for components of the Enterprise Federation provides a roadmap for orchestrating the interaction of the systems, which is frequently driven by human users. These best practices for orchestrating the interaction of the data and systems are still being developed.

Procedural: As much as practical, the orchestration of the interaction among these systems will be automated. Again, because this level of abstractions is the newest, the development of orchestrations algorithms is in its infancy.

Evolution: The orchestration of the interaction among the members of the Enterprise Federation is evolving as the different historical contingencies of the various services (Air Force, Army, and Navy) are incorporated and as the differing enterprise systems for GIS and real property asset management are included. The orchestration techniques must be allowed to evolve with some independence as each service meets its different mission needs.

5.5.2 Desktop Federation

The desktop federation of PAVER and PCASE has been hosted on a common pavement engineering desktop for the past fifteen years. Users can install either program alone or install both to get a seamless integration of the tools.

5.5.2.1 Ontologies

Declarative: The integration effort began with many months of matching up key concepts in the systems in order to arrive at a shared ontology for the shared inventory elements. The shared ontology contained Network, Branch, and Section from PAVER as well as the pavement use and surface type. The notions of non-destructive test data and layer definition were taken from PCASE. Additional ontological elements were identified as predominantly being associated with one system but these concepts are useful, in principle, to both.

Procedural: Each of the two members of this federation had its own database structures behind the ontologies. Once the competing ontologies were reconciled, we needed to unify the persistence while respecting the constraint of allowing each system to retain its separate identity and independent development team. We accomplished this by having the system construct a single logically unified database by linking the various databases from PAVER and PCASE into it. In this way, we could manage a unified common set of core inventory data and allow for the union of additional persistent data, which is managed by one member of the federation but could be reported by either.

Evolution: The PAVER and PCASE teams have been free to control the portion of the ontology, which is predominantly controlled by one group. Changes to the core shared inventory elements required both groups. At one point, PAVER had two different versions (versions 5 and 6) deployed with different file formats and internal object structures. Backward compatibility with PCASE was provided by use of a façade, which made version 6 appear to provide the same ontological model as version 5 to PCASE.

5.5.2.2 Discovery

Declarative: The desktop federation of PAVER and PCASE might be comprised of either application alone or both together. We honored this constraint by having the Desktop executable search for a file with a special extension so as to discover what tools were available. These files described the root level application object to instantiate and put into a collection of application objects being hosted by the desktop. In PAVER versions 5 and 6 and PCASE versions 2.08 and 2.09, the menu system for each application was also declaratively stated in a tabular format and discovered by the desktop load on startup.

Procedural: PAVER versions 5, 6, and 7 all use a search to discover applications to load. Starting with Version 7, the menu system has been described procedurally and merged on desktop startup. The Version 7 desktop can also be used to host additional tools, which are derived from PAVER, i.e. the Image Inspector (for analyzing roadway images) and the Field Inspector (for collecting distress data in the field) both use this same desktop together with a procedural menu system and a configuration file to be discovered by the desktop when it loads.

Evolution: For the past fifteen years, the discovery mechanism in each version of the shared pavement management desktop has supported the independent evolution of PAVER and PCASE and, more recently, plugins.

5.5.2.3 Composition

Declarative: The shared ontology dictates that both PAVER and PCASE have a shared inventory, which is a composition of inventory levels. The shared ontology also defines time series data, which are items with a date under the sections, e.g. Work History, Inspections, and Conditions.

Procedural: In addition to the declarative (predefined) compositions, users and tools can add members to the collection of "condition measures." PCASE adds measures and users may add their own measure scales. When additional condition measures are added, it is as if there are now more compositions of conditions available for data entry and reporting. The GIS reporting tools and the tabular reporting tools will detect the new compositions of condition data and show the user reports with the new types of data.

Evolution: The available condition measures have evolved with the evolution of best practices and the advent of more automated data collection tools. In addition, users can control the presence of certain compositions of descriptive fields, which are used for asset management and even hide some of the compositions or repurpose them for another use in the system. In this way, the system supports a limited amount of evolution of the composition of the data. The advent of "Add In" modules in version 5, which have been expanded in version 7.1, brought the ability to add procedural behavior to these new compositions of data and to compose behaviors for new compositions through procedures encoded in a dynamically loaded plugin module.

5.5.2.4   Orchestration

Declarative: We say that the Desktop "orchestrates" the separate application objects (PAVER and PCASE) because it handles the process of wiring up the objects and passing action messages from menu items and such to the correct application object. This declarative definition of the orchestration behavior is enforced by interface contracts, which an application object must honor. Starting in version 7, the declarative machinery was enhanced to allow application objects and individual component tools within an application object, to register "interest" in one or more events and to be notified when they occur.

  Procedural: The shared desktop orchestrates some activities with a combination of procedural processes and notifications, e.g. when a user changes unit system (e.g. Metric to English) or changes language (e.g. from English to Italian), the desktop can handle some of these operations itself and must orchestrate the notification and update process for all of the tools within all the applications and plugins. In versions 5 and 6, notifications were broadcast to all participant tool objects independent of declared interest in the changes. In version 7, tool objects can register to receive notifications and are responsible for handling them without further help from the desktop. This change of architecture was needed to allow the same mechanism to be used for Windows forms and web pages, which realize their user interfaces in radically different ways but can share this orchestration logic.

  Evolution: As mentioned before, the desktop orchestration machinery has been required to evolve in order to handle PCASE 2.08 and PCASE 2.09 together with versions 5 and 6 of PAVER. More recently, the same family of orchestration tools has been used to control a version of PAVER 7 for Windows and the Web. Finally, the orchestration machinery for notifications about a user's focus of interest is open ended and can be extended by tools at runtime to support notification messages, which were not originally predefined.

5.5.3   Tool Federation

The Tool Federation represents a subdivision of the PAVER level of abstraction into a collection of partially separable tools. In version 7, the system was extended to support multiple users in either a web or thick client—server configuration. The client-server interaction is through web services using Microsoft's Windows Communication Framework (WCF). Unlike many stateless web service protocols, WCF supports the notion of a user session and we have extended that notion to allow us to have web services, which can initialize web services with a user specific data context and return what amounts to a handle to the web service.

  Web service protocols are often stateless and each service call must stand on its own. In our systems, we often need to have what amounts to a web service that can return a handle to another service where the new service shares common user data. We use the WCF notion of a user session to implement what amounts to web services that can return web service handles in order to support complex chains of processes. In addition to the separate web service tools within PAVER, the Tool Federation also includes the "Add Ins" (plugins) to the desktop, which can also extend PAVER. These additional modules live almost entirely on the client, but they have access to the persistent data on the server through a constrained interface. Because these tools live within the level of the overall PAVER system, they have access to many of the SOA elements of the overall application. In this section, we discuss only those aspects that are specific to the Tool Federation.

5.5.3.1   Ontologies

Declarative: The WCF protocol makes the distinction between declarative ontologies (Data Contracts) and Procedural Ontologies (Operation Contracts). These data contracts are not interesting for this analysis because they are more an artifact of technology than a true ontology. Our preference for developing models of the application domain and then solving problems using the model makes these data contracts more like ontologies than they would otherwise be. The declarative ontologies for plugins are directly related to this material. Version 71, extends the machinery for plugins to be able to store data in the main inventory database as well as the separate and potentially shared system tables database. These data can participate in the Import/Export process for sharing data between users by file sharing and updates to these data are multiuser safe. The Desktop imposes an overall declarative structure on these data, which involves a GUID for the plugin and a "Class Name," which must be supplied by the plugin when retrieving or storing data. The "Class Name" is basically like the name of a component ontology, which is managed by the plugin. This block of data can consist of a miniature database, which is composed of a collection of potentially interrelated tables. The plugin provides the meaning for the internals of these tables and the Desktop imposes the structure of the key fields used to distinguish the individual blocks of data.

    Procedural: Again, the procedural ontologies associated with the WCF services are needed in order to break PAVER into a set of component web services, but they are relatively uninteresting for this work. The plugin machinery provides all of the aspects used by WCF and several more. The procedural ontologies are defined by a special interface contract DLL component to which plugins must conform. The various interface contracts define how the desktop and the plugins can communicate. For example, a plugin can place menu items at various locations in the system (main menu items or GIS map display options) and the procedural ontology for plugins governs how this is done. In addition, the plugin itself controls the semantic meaning of the items and their associations with other items through its algorithms for data collection, validation, value added processing, and presentation.

    Evolution: The plugin machinery provides a means for ontological evolution by its mere presence. The high level constraints of the storage of mini-databases only modestly constrain how these items are packaged. In some situation, such as for non-destructive testing, there are domain ontologies, which can be supported by plugins even when these ontologies vary from one equipment vendor to another.

5.5.3.2 Discovery

Declarative: The Desktop uses declarative information about an inventory, which the user can access through File/Open, to determine whether to access data locally or on a remote server. The discovery process for plugins is more interesting for this analysis. Each user has a declarative (tabular) set of preferences, which specifies whether an available plugin should be loaded for that user or not. The desktop manages the discovery of installed plugin modules on startup and handles the dialog with the user to discover which items the user wants to access. The selection of plugins to use is persistent between user sessions and is user specific.

Procedural: The Desktop uses a procedural search together with requirements for contracts to be implemented by plugins to discover available plugins on startup. For software security reasons, this discovery process is constrained to distinguish between "certified" plugins and "external" plugins and is constrained to a tightly controlled set of locations. The underlying procedural discovery and composition of plugins is capable of supporting less restrictive policies. Perhaps the most interesting discovery operation involves plugins being able to discover other plugins. When a plugin is loaded into the collection of active plugins, the desktop can determine if the plugin is willing to be visible to other plugins. A plugin with a dependency on another plugin can determine if the other plugin has been loaded and can obtain a pointer to the other plugin. After the desktop has facilitated the connection between the plugins, they may continue to cooperate directly or indirectly through the database and/or the user interface.

Evolution: The plugin machinery exists for the purpose of allowing the discovery of incrementally added modules of functionality. Thus, its primary purpose is to allow the main desktop system to evolve separately from these additional tools. Although the discovery of one plugin by another plugin is mediated by the desktop, there is relatively little constraint on how these interactions are handled once established.

5.5.3.3 Composition

Declarative: The WCF composition of web services in the main system reduces the complexity of the overall system and provides for incremental recovery from failed client—server interactions. Increasingly, new features in the main system are being packages as "certified" plugins in order to simplify the extension of the main system and to simplify the system versioning problem. These certified plugins are functionally part of the main application, but they benefit from many of the simplifications resulting from loosely coupled SOA modules. Finally, the user can use the desktop plugin manager to select or deselect a plugin for inclusion in the user interface.

This selection process can be thought of as a (tool) composition process, which is controlled by the user. Some users may not need to use all of the functions in the system and this dynamic tool composition allows for a simplified user interface.

Procedural: As described already, the plugin machinery supports mechanisms for dynamic composition of algorithms and data. One commonly occurring example of procedural composition involves reader modules for field data, where there are competing vendor file formats, e.g. falling weight deflection data. The field data may be in any of a dozen data formats and each vendor will usually have the best reader for its format among competing readers. In this situation, we have been forced to allow plugins to be able to specify a precedence ordering over potential tool compositions to favor the vendor's reader over those of others. By supporting the composition of an ordered collection of readers, plugins can share a common set of reader code rather than needing to maintain separate versions of what should be substantially the same functionality.

Evolution: Again, the plugin machinery exists to support evolution of incremental parts of the system without a need to version the overall system. Also, the composition of algorithms and data from multiple sources (i.e. separate civil engineering firms) allows each plugin to leverage the shared inventory and reporting tools of the desktop.

5.5.3.4   Orchestration

Declarative: The desktop allows plugins to supply call back objects that implement required interface contracts for use when a user requests functionality from a plugin item. The desktop manages the user interface for the invocation of plugin functionality as well as loading an initializing the plugin. Also, plugins that plan to offer functionality to other plugins will usually define one or more interface contracts to facilitate the interoperation of the plugins. The desktop orchestrates the connection between the plugins and may not be involved going forward.

Procedural: Because these are decision support tools for practicing pavement engineers, a portion of the orchestration of the component services is actually done by the human user when implementing best practices in the domain. We may say that the human user helps to orchestrate the operation of the various services through the sequence of engineering tools used and through the various tool windows concurrently opened. The user interface supports the concurrent use of a collection of component tool windows, which may have originated from the main application or a plugin. The desktop also supports limited abilities to remember tool window configurations from one user session to another, which allows each user to control some of the orchestration settings across sessions, e.g. a user may request that a GIS map be opened on startup if available and the locations and sizes of tool windows can be remembered, which allows the user to play a role in the orchestration of tool windows.

Evolution: The evolution of the orchestration of the interaction of tools comes in part from the ability to add new tools with a variable number of user interface elements for launching components of each tool or plugin. In addition, the ability of plugins to interoperate after they have been connected to each other by the desktop provides a path for the inclusion of new orchestration patterns by the inclusion of additional plugins. In fact, we are planning to mitigate the client versioning problem by the use of more and more certified plugins rather than main system versions. Because the orchestration of the underlying engineering algorithms depends, in part, on the user's selection of sequences of engineering operations and/or reporting parameters, there is room for the orchestration of the tools by the combination of the human user and the desktop to evolve with changing best practices.

5.6 Fractal Issues We Have Identified

Some of these issues were known to the authors prior to encountering them in the pavement management domain and some of them have emerged during the course of working in this domain. In this section, we will summarize the high level issues we found and relate them back to the real world example. We have identified the following overarching principles from our experience:

- Finiteness limits drive the need for increased structure—constraints on time and/or resources can require a more highly structured solution in order to solve the problem at hand within the given constraints;
- SOA Federations favor some structural patterns over others—some patterns of software structure produce better SOA Federations than others and there are some guiding principles for choice of good patterns;
- SOA Federations favor late binding—late binding to a particular solution element avoids the details of selecting the best component tool until it is actually needed;
- Mixed Initiative Dialog—a recognition that SOA systems, whether made up of human or machine actors interact more as peers and either side of the communication may have the initiative from time to time;
- Trust, Reliability, Ability, and Authoritative Source—when designing SOA systems or discovering a candidate for dynamic composition, potential human or software actors cannot always be treated equally.

For each of these issues, we provide a theoretical basis and a basis in our experience. These two items independently provide support for our conclusions about the fractal nature of these issues and their importance in SOA.

5.6.1 Finiteness Limits Drive the Need for Structure

Baskin et al. have explored this idea in detail in [32]. In the interest of keeping this material self-contained, we briefly summarize the key parts of this principle here.

Theoretical Basis: Think of a problem to solve in the form of a mathematical function, which is a subset of the domain of input values cross the range of output values. Any computable function can be represented by a universal Turing machine with a starting tape containing a definition of the finite state controller for the machine and a finite amount of starting data. The Kolmogorov complexity [33] of the function can be thought of as the length of the shortest universal Turing machine starting tape that solves the problem. Multiple minimal starting tape can exist with different structures. The size of the function can be taken to be the length of a starting tape with a simple finite state controller that exhaustively searches a table of input/output pairs and matches the given input value(s) to find the corresponding output value. If the size of the function is equal to the Kolmogorov complexity, then there is no room for the use of structure within the function to reduce the size of the starting tape and still solve the function. If the Kolmogorov complexity is less than the size of the problem, then there are some combinations of the various input values that lead to a common outcome. In this formulation, we can think of finiteness limits as limits on the run time of the universal Turing machine and/or limits on the length of the starting tape (i.e. limits on the size of the starting data and/or the state transition diagram for the finite state machine controller).

When finiteness limits are imposed, the nature of the solution must be modified from an exhaustive lookup table to exploit the commonality among subsets of the presenting inputs. This process innately involves the explicit incorporation of these patterns of commonality into the structure of the solution. The recognition of the commonality in the structure of the problem requires increased structure within the solution and that structure must exploit the patterns of commonality within the function. Finiteness limits may be expanded also. When finiteness limits are expanded, e.g. by increasing speed and/or expressive power of solution platforms, then either less highly structured solutions can be used or more complex functions can now be addressed within the expanded limits.

Domain Examples: The initial impetus for the integration of PAVER and PCASE into a single desktop came from the imposition of a finiteness limit, i.e. the push to replace two (partially redundant) inventory definitions with a single (unified) inventory structure. In a similar way, combining the functionality of these closely related tools matched the reduction in staffing (another finiteness limit), which accompanied the reduction in field engineering office head counts during the past twenty years. The more recent pursuit of enterprise level asset data management was stimulated by a DoD push to respond to pressure on budgets over the past decade and a need to be more efficient in the allocation of resources. The push for plugin modules in the Desktop represents an attempt to reduce the cost (a finiteness limit) of civil engineering tools by leveraging the common inventory and presentation tools so that each vendor's civil engineering tool does not need to have its own (redundantly expensive) version of these shared tools. In fact, it might be possible for some civil engineering tools to actually have a shorter development time, lower development cost, and increased range of functionality by leveraging the desktop tools. In order to exploit these things, the complexity of the plugin tools must generally be increased by the need to conform to the structures demanded by the plugin framework, which is a layer of abstraction that could otherwise be avoided.

5.6.2 SOA Federations Favor Some Structural Patterns Over Others

As we saw in the previous section, some representations of the form of the solution can embody more knowledge about the structure of the problem at hand than others. Among competing structures for solutions, we have found that some structures are more useful than others. We have identified the following structural patterns, which are especially useful for constructing SOA federations, but are also good software engineering principles as well:

- Matching—the patterns of coupling and functional decomposition in the solution will be simplest and more robust if they match the analogous patterns in the problem domain itself;
- Favor Composition over Specialization—is a common adage in software engineering but it is especially useful in SOA federations because this bias makes discover and dynamic composition much easier;
- Manage Variation Explicitly—try to find a balance point between exhaustive enumeration of standards and the chaos that results from a lack of attention to the explicit management of patterns of variation;
- Manage patterns of coupling to maximize convergence of the SOA solution under change—the evolution of best practices means that large SOA systems will change and convergence under change is essential;

Theoretical Basis: The imposition of finiteness limits drives the increase in structure for solutions. We first encountered the notion of matching in biology [34] but we have actively employed it for decades now. Among competing structures for solutions, biology and, by analogy, best software practice, favors solutions whose structure matches the patterns of structure in the environment and in the structure of the problem. In software this means an analysis model where all object class names and relationships are recognizable by a domain expert as a model of the domain. The domain model will be more stable than any particular solution structure and will be better suited for evolution to solve related problems later. The principle of matching suggests that wherever there are components of the problem domain with differing rates of evolution, there should be a factor point (composition) to allow two components of the software to evolve separately. Similarly, when two components in the problem domain are highly coupled, then they should be coupled in the structure of the solution.

The virtue of composition as a tool for modular replaceable parts is well known. The notions of Discovery and Composition in SOA are intended to directly exploit composition. Composition is also a key tool for avoiding duplication of functionality in multiple places where the evolution of the functionality needs to be shared.

Explicitly managing variation is closely related to the principle of matching, which was described earlier. It involves finding boundaries in the domain where elements can vary separately and then match that boundary with a comparable software module boundary. Either enumerate the variety completely (e.g. Metric or English unit systems) or explicitly allow for variation in a constrained way, e.g. using in interface contract. Patterns in dynamical systems can also be shown to suggest that some structural patterns support evolution better than others [35].

First derived in mechanical engineering but also applied to software, the principles of Axiomatic Design [36] show that it is possible to form a dependency matrix describing patterns of coupling among software modules and describing the pattern of use of software modules to solve a problem. When the dependency matrix for the software modules can be made lower triangular, then there is a precedence ordering over the modules such that they are stably convergent under changes in them. Software modules and patterns of using them to solve a problem for which the coupling matrix cannot be made lower triangular require simultaneous and coupled changes in multiple places and are not convergent under change. Changes in software modules can be driven by changes in the requirements and changes in technology. Explicitly managing patterns of coupling facilitates software evolution.

Domain Examples: The object model for PAVER is a domain model, which is then used implement the various requirements for best practices. It matches the real world pavement domain and, hence, has evolved well. The origin of the enterprise systems as separate "smokestacks" also mirror the substantial separation of these functions. The origin PAVER as a separate system made it natural to see the engineering rules and analysis modules as an extension of the basic inventory. In hindsight, this use of specialization to add functions to the basic inventory was a mistake. We should have favored composition and, then, when PCASE came along, both PAVER and PCASE could have shared the inventory as a commonly held part. Unfortunately, it has not proven practical to fix this mistake and a workaround has been required. Had we favored composition originally even when there was not an obvious use for it, we would have had a better domain model and an easier time integrating PCASE with a shared inventory.

Managing variation explicitly has been done extensively in PAVER where certain things can be locked down via analytic closure (e.g. surfaces are flexible, rigid, or unpaved). In other places where the domain allows for meaningful variation and/or extension, then users are allowed to extend built-in types and are required to supply engineering attributes of these types so that they can be used by the analysis algorithms. The most interesting examples of explicitly managing variation come from the Enterprise Federation. The integration of GIS attributes between PAVER and the enterprise GIS attributes has been seen as a problem of locking down the attributes to be supplied by PAVER to the enterprise GIS system.

At first glance, a specifically enumerated set of attribute values would appear to make the problem simpler but it actually makes it harder. Especially in a GIS where users can define their own coloring strategies for attributes, there is constant demand for more and varied attributes from PAVER. The ultimate solution was to define a GUID in the GIS to be matched to a GUID in PAVER and then to have the GIS user "join" the GIS data to the PAVER data as needed. This explicit management of variation by identifying the only legitimate standardization and tolerating complete freedom after that point is simultaneously an instance of matching, explicit management of variation, and late binding as presented in the next section.

The integration of the real property/asset management data also provides an instance of matching, explicit management of variation, patterns of coupling and late binding. Both the pavement domain and the real property domain have an inventory hierarchy and the hierarchies are highly correlated. The original software requirements given to the PAVER development team were to modify PAVER to enforce the congruence of the pavement domain inventory hierarchy and the real property hierarchy. The high correlation of these two inventory structures (80 %) made this convergence appear to be simpler than allowing them to be different. By applying principles of domain modeling and matching, the PAVER development team was able to push back and get permission to implement the real property tags at a lower level than originally requested with the justification that there were legitimate domain rationales for an item to be at two incompatible places in the two hierarchies and by tagging lower level elements, it was possible to dynamically roll up the PAVER inventory according to tags representing a somewhat different composition. The system allowed tags to be supplied at the originally requested higher level and only exceptions to that assignment were needed at the lower level. This approach exactly matched the predicted pattern of domain variability. Despite repeated attempts to force field engineers to use the original exact correspondence and tag at the originally requested level, the system was not usable for some locations. Once users were allowed to tag exceptions at the lower level, the system was accepted. This is an example of predicting a domain requirement based upon fractal SOA principles and persevering in the face of resistance from contract monitors.

Explicit management of patterns of coupling is a combination of software development practices and domain modeling. It is possible to explicitly manage patterns of coupling by using a heavily constrained N-tier software development model. Each tier consists of an interface contracts module, which is visible only to the layer above, and an implementation module, which is not visible to the layer above. These constraints can be enforced by allowable patterns of reference among modules. Using this approach, it is not possible for implementation code in one layer to become coupled to implementation code in a lower layer. Using a strongly contract driven pattern of allowable coupling does lead to a larger number of modules than would otherwise be required. If needed, these decoupled modules can be merged at the last minute to ease packaging while retaining the constraints on coupling.

5.6.3 SOA Federations Favor Late Binding

As shown in the issues surrounding trust, the SOA ideas of discovery and composition imply a substantial departure from the historical notions of "link editing" all of the software modules in a software system into a single executable at load time. Like other issues we have identified, the notion of late binding applies at all levels of abstraction.

Theoretical Basis: The notion of late binding has actually been around for a long time. In the original LISP implementations, the boundary between program and data was blurred and a LISP program could build a list of instructions and then execute those instructions! We stop short of that ultimate example of late binding in our discussions here, but we note that the inclusion of human users as one of the SOA Federation members, we achieve late binding of a similar order because the human users may elect to compose data sources in entirely new ways and may interpretively execute algorithms by invocation of SOA Services where the algorithm exists only inside the user's mind or is codified in a best practices manual.

We find another model for late binding in what Lu et al. [37] has called Engineering as Collaborative Negotiation (ECN). The ECN paradigm was developed for mechanical engineering product design and we have applied some of its principles in our work. The basic idea of ECN is to identify constraints on the design result as early as possible but with the broadest tolerances possible. This approach is in contrast to more traditional mechanical engineering design approaches that emphasize the preparation of relatively specific designs for major subsystems as the overall product design matures. By identifying broad constraints on the design as early as possible and delaying selection of specific design values until as late as possible, designers can detect conflicts in the design much earlier than would otherwise be possible. This late binding approach leads to a more agile and cost effective design process. We find a similar situation in SOA Federations because the delay in binding to details can afford opportunities for opportunistic selection of tools for composition.

Domain Examples: The mechanisms for late binding are different at each level of abstraction, but there is value to late binding at each level. The matchup for GIS data and Real property data needs to be late bound because the Air Force, Navy, and Army all use PAVER but use different enterprise systems. We can use a single data harvest mechanism, but each service must use its own tools for accessing the data. Secondly, the data may be referenced in place (planned for the Air Force) or by accessing a "published" copy of data (Army and Navy).

The Desktop federation of PAVER and PCASE uses late binding when importing data and when exporting data. This federation has a collection of application tools, which may contain one tool or both tools depending upon what the user has installed. The import and export tools actually bind to the collection of applications for putting data into an export file or bringing in data from an import file. These operations are internalized into each of the Desktop SOA federation members. The data for all plugin modules is processed by the desktop for import/export but each plugin module will only be bound to data if there is both data in the plugin persistence and the plugin is activated for the user. Another example of late binding occurs when the user opens a database. Any given database might have been created with PAVER alone, PCASE alone, or both PAVER and PCASE installed. During the file open process, each application is allowed the opportunity to create any missing databases, which might be exclusively managed by one tool and to add them to the single logically unified database. In this way, the number of databases and data tables is late bound at the time of file open.

Late binding is a fundamental part of the plugin machinery. Some plugin modules may be installed as an integral but optional module. We refer to these plugin modules as "certified" because they can be recognized as being compiled with the same trust level as the main application code. The user can opt to include these tools in the user interface and, thereby, cause a late binding of user visible functionality. For plugin modules from other civil engineering firms, there is both late binding of those tools to the desktop and also there can be late binding of these tools to each other through a SOA discovery and composition protocol, which is mediated by the Desktop as it orchestrates the initial setup of each plugin module.

5.6.4 SOA Federations Contain Mixed Initiative Dialogs

One reason that we have used the word federation throughout this discussion is the realization that we are bringing together a collection of relatively coequal participants where the union of what the participants know and can do is needed to solve complex problems. This relative symmetry of the participants gives rise to the situation where one federation member may predominate at one point in time and another participant at another time, and, hence, gives rise to the need to support a mixed initiative dialog at all levels of abstraction.

Theoretical Basis: We borrow the notion of a mixed initiative dialog from intelligent tutoring systems [38, 39] in which two semi-autonomous agents interact with each other while exchanging the control over the interaction. In intelligent tutoring systems, the human student is made to be more actively involved by being put into a position of active participation rather than passive listening. In our work on engineering field office automation in the 80's, we saw a clear role for supporting the field engineer with individual decision support systems and later groups of engineers with group decision support systems. By keeping the practicing engineer involved in the decision making process, some of the harder problems can be offloaded onto the human user and, thus, we arrive at expert support systems (where the expert is the human decision maker) rather than expert systems (where the computer system is expected to be an expert).

Domain Examples: In our civil engineering maintenance management tools, we emphasize supporting the decisions of a practicing pavement engineer rather than replacing their decision making with expert-derived rules or algorithms. Examples of this include the provision of common domain defaults for all required inputs for work planning together with the ability for the engineer to override the defaults or extend them, e.g. an initial set of work types, surface types, cost tables, and budgets. The incorporation of plugin modules is another example of a mixed initiative dialog where in the installation of a plugin module requires administrative privileges but does not automatically activate the plugin for the user. By using the "add in manager" tool, the user can control which Add Ins (plugins) are actually presented in the user interface. Finally, the composition of functionality among different plugins requires a dialog between the desktop and the plugin to expose it to other modules, which, in turn, enter into another mixed initiative dialog to exchange data and/or provide composite calculations.

At the Enterprise SOA Federation level, we see a mixed initiative dialog between the systems in the federation where the authoritative source (Spatial Data Standard system or real property/asset system) might temporarily give way to an operational authoritative source like PAVER because a field engineer standing on the pavement may be a more trusted source for those data as a byproduct of the field data collection.

Within the SOA Federation containing PAVER and PCASE, we again see a mixed initiative dialog, which we are still trying to fully realize: one system indicates that the most effective intervention is reconstruction and the other designs the details of the new construction. Each member of this federation draws upon the shared inventory but has the initiative for capturing and processing largely disjoint sets of time series data and analyses.

While there is technically a mixed initiative dialog between the various web-services for PAVER itself, the late binding of add on modules is more interesting and subsumes these internal issues. The mixed initiative dialog begins with the user electing to use the "Add In Manager" tool to activate one or more installed additional modules. Each plugin module has the option at this time to require a license key from the user and either it agrees to become activated or not. Once the plugin module has been activated, it can request that the Desktop host user interface items by which the user can request the plugin to respond. The Desktop and the human user have the initiative more often going forward, but a plugin module can do things like monitor a GPS feed and/or host its own user interface and the result of these data can be pushed back into the Desktop Federation to shift the user's focus of attention to see the newly selected data. This bi-directional control is most visible when there are two GIS maps being synchronized bi-directionally—one from PAVER and one from a plugin. To the human user, the two maps are part of a single unified interaction experience but the communication among the SOA Tool Federation members is not a seamless as it appears to the user.

Finally, plugin modules can call upon other plugin modules in something approximating the normal SOA Discovery and Composition techniques. During the instantiation of the plugin module, the Desktop and the plugin engage in a back and forth dialog by which plugin modules can agree to allow them to be used by others and a link between the modules can be directly established. Plugin modules that communicate directly with each other no longer need the Desktop to mediate further communication. Because these modules may also have their own user interaction items, they behave more like co-routines than procedure call services and, again, we find a mixed initiative dialog between the plugin modules.

5.6.5 SOA Federations Depend upon the Explicit Management of Trust, Reliability, and Authoritative Source

As the length of this section title suggests, management of the issues surrounding trust is a somewhat messy problem. It involves issues between humans only, between software modules only, and between software systems and human users of those systems. As we will see below, this is also an issue between software development teams whose software systems will be members of a SOA Federation.

Theoretical Basis: The distinction between Authentication (do I believe you are who you claim to be) and Authorization (what I allow that authenticated identity to do) is well established, and we build upon that as a foundation for related issues. Implicit in the SOA notions of Discovery and Composition is the ability to select among competing sources for providing a required service based upon things like performance and competence. Rephrasing this in terms of trust, we get the question: Can I trust you to provide correct services/data in a timely fashion? A subtly related point can be used to limit which features are shown to human users: Can I trust you to be able to understand this feature and not be overwhelmed by too many features? Both of these two questions can be thought of as complementary aspects of the notion of competence—competence to provide and to consume.

Reliability closely relates to competence: (1) Can I rely upon you to provide good data/services in a timely fashion? (2) Can I rely upon you to understand and not corrupt the data and services I expose to you? These questions apply equally between all SOA Federation members whether they be a human or a computer software modules.

During the work on the enterprise federation of civil engineering tools for DoD, we encountered the notion of "authoritative source," which means the agency and/or software system designated as the official "go to" source for a body or data and/or expertise. We have coined the notion of "operational authoritative source," which means a source different from the officially designated source but, at least temporarily, better able to supply reliable data at a particular place and time. As we will see in the domain examples, these two competing sources will be separates SOA Federation members and they will enter into a mixed initiative dialog whereby the authoritative source may be updated by the transient activity of the operational source.

Although we are still trying to fully understand the fractal nature of trust issues, we have identified a key role for trust in the following substantially separable areas:

- Trust between software development teams for different SOA Federation members (federates), which involves territoriality, use of tools you cannot control, schedule compatibility, and competence;
- Trust between federates and their sources, which involves authoritative sources, perceived competence/timeliness of the services, and dependence on services whose availability cannot be guaranteed;
- Trust in the longevity of the available SOA services, which is critical in civil engineering data management where data must be kept over decades and the nominal life of the asset may be 50–100 years;
- Trust in user competence, which causes tension between those managing low level data and managers who must necessarily look across data from multiple sites/facilities and may not understand low level data.

These diverse aspects of the concept of trust can be found in surprisingly diverse situations and frequently prove to be deceptively simple to identify and incredibly difficult to resolve.

Domain Examples: The designation of an authoritative source is entirely external to the issues being discussed here but the existing authoritative sources for GIS data and Real Property/Asset data constitute a constraint on major aspects of the Enterprise SOA Federation. Although these systems are the authoritative source, PAVER can frequently be asked to provide updated data of higher quality as a byproduct of field surveys. In fact, many field surveys using PAVER are now required to obtain the latest GIS and Real Property data and for use in the inspection process and to return for potential update of the authoritative sources. This mixed initiative dialog between the systems may be fully or partially automated, but the authoritative source is responsible for the eventual integrity of the data and may refine or refuse proposed updates. At the enterprise level, trust between software development groups and SOA Federation members tends to be resolved by designation of by a single shared authority, i.e. the Secretary of Defense. The existence of a single "owner" for all of the SOA Federation members is on a panacea for resolving trust issues but it is a surprisingly underappreciated necessary condition. At the enterprise level, the issue of trust for the competence of the various SOA Federation members has been resolved by their separate evolution and individual validation. Again, the existence of a common governing authority with control over allocation of resources and the independently justified existence of the systems means that the longevity problem is also solved.

The development of the original Federation of PAVER and PCASE, which predated the inclusion of SOA principles, involved a process of reconciling the respective pavement engineering domain ontologies and establishing trust between the separate development groups (one made up of government employees for PCASE and the PAVER development team). Because there is a natural process of turnover in software teams, and because these civil engineering systems deal with problems over a period of decades, it has proven surprisingly difficult to maintain this trust between the groups over time.

The trust issues between plugins provide the richest examples of trust issues in the three layers of abstraction. When plugin modules can originate with the PAVER development team or with competing civil engineering firms there can be no illusion of a common controlling authority and no illusion that these SOA Federation members will always be there because plugin modules may be separately developed, licensed and distributed. Increasingly, these plugin modules depend upon external web services for data and operations and, again, continuity of access is qualitatively lower than for modules distributed as part of the main system.

The issue of trust is not an absolute distinction. An external authority can designate an authoritative source, but that distinction is artificial and external to all of the issues of the SOA Federation. The trust issues generally exist on a precedence ordering and not as a crisp distinction. We can illustrate this point with the following detailed example.

5.7   Conclusions

This chapter summarizes insights gained from more than twenty years of software development, maintenance, and evolution of a major pavement management system (PAVER™). We consider the traditional SOA concepts: (1) Ontologies, (2) Discovery, (3) Composition, and (4) Orchestration. Often, discussions of SOA techniques focus on stateless operations, which certainly have their place. Managing the persistence of time series data is essential whether orchestrating the collaboration of human civil engineers in a technology mediated federation or managing diet/exercise data with an app on a cell phone. Accordingly, we have cited this as a cross-cutting concern. We conclude that time series trends in such data are much more meaningful than any single snapshot of data. This observation leads us to an additional dimension, which cuts across all of the SOA concepts above: Algorithms versus Persistent State. Finally, during twenty years of experience in the pavement management domain, we have become attuned to the issue of evolution of best practices and associated decision support software systems, as well as the need for SOA Federations to support this evolution. This fact gives rise to a third dimension which we have explored in this work: Evolution. After reviewing our experiences with SOA Federations at three levels of abstraction, we have found the following basic principles to be self-similar at three levels of abstraction:

- Finiteness limits on time, participants, and/or resources demand more highly structured solutions;
- SOA Federations work best when they (a) match patterns of coupling/evolution in the domain, (b) favor composition, (c) manage variation as a first class issue, and (d) explicitly manage patterns of coupling;
- SOA Federations benefit greatly from late binding—especially when paired with management of variation;

- SOA Federations work best when there is a mixed initiative dialog among federates and human users;
- Trust issues must be managed by SOA Federations and among software development teams.

We have found all of these fractal principles in our historical review and we have used them prospectively to guide the development of new SOA system federates to good effect.

References

1. Shahin, M.Y.: Pavement Management for Airports, Roads, and Parking Lots. Chapman & Hall, New York (1994)
2. Reinke, R., et al.: Domain frameworks for collaborative systems: lessons learned from engineering maintenance management. CTS **2007**, 396–405 (2007). doi:10.1109/CTS.2007. 4621780
3. Zdun, U.: Pattern-based design of a service-oriented middleware for remote object federations. ACM Trans. Intern. Tech. **8**, 3, Article 15 (2008). doi:10.1145/1361186.1361191
4. Li, Z., Cai, W., Turner, S.J., Pan, K.: Federate migration in a service oriented HLA RTI. 11th IEEE Symposium on Distributed Simulation and Real-Time Application, pp. 113–121. doi:10. 1109/DS-RT.2007.31
5. Wang, W., Yu, W., Li, Q, Wang, W., Liu, X.: Service-oriented high level architecture. In: Proceedings of summer computer simulation conference, 2008. Article 16
6. IEEE: Standard 1516 (HLA Rules), 1516.1 (Federate Interface Specification) and 1516.2 (Object Model Template), September 2000
7. WSDL: Web services description language (WSDL) Version 2.0 Part 1: Core Language http:// www.w3.org/TR/wsdl20/. Accessed 20 Mar 2014
8. SOAP: SOAP Version 1.2 Part 0: Primer (Second Edition) http://www.w3.org/TR/2007/REC-soap12-part0-20070427/
9. XML Schema: XML Schema Part 1: Structures Second Edition http://www.w3.org/TR/ xmlschema-1/. Accessed 20 June 32013
10. Seo, C., Zeigler, B.P.: Simulation model standardization through web services: interoperation and federation on the DEVS/SOA platform. In: Proceedings of symposium on theory of modeling and simulation—DEVS integrative M&S symposium, 2012. Article 46
11. Li, J., Karp, A.H.: Access control for the services oriented architecture. In: Proceedings of ACM workshop on secure web services, pp. 9–17 (2007). doi:10.1145/1214418.1314421
12. specs@openid.net. "OpenID Authentication 2.0 Final." 2007. Available online at http:// openid.net/developers/specs/
13. Liberty Alliance Project: Liberty ID-WSF web services framework overview. Version 1.1, 2005. Available online at http://www.projectliberty.org/liberty/specifications__1
14. OASIS: Web services security: WS-security core specification 1.1. OASIS Standard, 2006. Available online at http://docs.oasis-open.org/wss/v1.1/
15. OASIS: Security assertion markup language (SAML) 2.0 Technical Overview, Working Draft 05', 10 May 2005. http://www.oasisopen.org/committees/download.php/12549/sstc-saml-techoverview-2%5B1%5D.0-draft-05.pdf
16. Thomas, I., Meinel, C.: An identity provider to manage reliable digital identities for SOA and the web. In: Proceedings of IDtrust '10, pp. 26–36 (2010). doi:10.1145/1750389.1750393
17. Hatameyama, M.: Federation proxy for cross domain identity federation. In: Proceedings of DIM '09, 13 November 2009, pp. 53–62. doi:10.1145/1655028.1655041

18. Anastasi, G.F., Carlini, E., Dazzi, P.: Smart cloud federation simulations with CloudSim. In: Proceedings of ORMACloud'13, June 17, 2013, pp. 9–16 (2013). doi:10.1145/2465823. 2465828

19. Al-Masri, E., Mahmoud, Q.H.: Identifying client goals for web service discovery. 2013 IEEE international conference on services computing 2009, pp. 202–209. doi:10.1109/SCC.2009.60

20. Dabrowski, M., Pacyna, P.: Cross-identifier domain discovery service for unrelated user identities. In: Proceedings of the 4th ACM workshop on digital identity management, pp. 81–88 (2008). doi:10.1145/1456424.1456438

21. Tolk, A., Turnitsa, C.D., Diallo, S.Y.: Model-based alignment and orchestration of heterogeneous homeland security applications enabling composition of system of systems. In: Henderson, S.G., Biller, B., Hsieh, M-H., Shortle, J., Tew, J.D., Barton, R.R. (eds.) IEEE winter simulation conference, Dec 2007, pp. 842–850. doi:10.1109/WSC.2007.4419680

22. Tolk, A., Diallo, S.Y., Turnitsa, C.D.: Mathematical models towards self-organizing formal federation languages based on conceptual models of information exchange capabilities. In: Mason, S.J., Hill, R.R., Mönch, L., Rose, O., Jefferson, T., Fowler, J.W. (eds.) IEEE winter simulation conference, Dec 2008, pp. 966–974. doi:10.1109/WSC.2008.4736163

23. Rathnam, T., Paredis, C.J.J.: Developing federation object models using ontologies. In: Ingalls, R.G., Rossetti, M.D., Smith, J.S., Peters, B.A. (eds.) Proceedings of the IEEE 2004 Winter Simulation Conference, pp. 1054–1062 (2004). doi:10.1109/WSC.2004.1371429

24. Calvanese, D., De Giacomo, G., Montali, M.: Foundations of data-aware process analysis: a database theory perspective. In: Proceedings of PODS '13, 22–27 June 2013. doi:10.1145/2463664.2467796

25. Reichert, M.: Process and data: two sides of the same coin? In Proceedings of the On the Move Confederated International Conference (OTM 2012), volume 7565 of Lecture Notes in Computer Science, 2–19 (2012)

26. Dobos, L., Csabai, I., Szalay, A.S., Budavári, T., Li, N.: Graywulf: a platform for federated scientific data and services. Proceedings of SSDBM '13, July 29–31 2013, Baltimore, MD, USA, 2013 ACM 978-1-4503-1921-8/13/07 (Pázmány Péter sétány)

27. Krizevnik, M., Juric, M.B.: Improved SOA persistence architectural model. ACM SIGSOFT Newsletter **35**(3), 1–8 (2010). doi:10.1145/1764810.1764821

28. Williams, K., Daniel, B.: An introduction to service data objects. Java Developer's J. (2004)

29. Carey, M.: The BEA AquaLogic Data Services Platform. Proceedings of SIGMOD 2006, June 27–29, 2006, Chicago, Illinois, USA. Copyright 2006 ACM 1-59593-256

30. Takatsuka, H., et al.: Design and implementation of rule-based framework for context-aware services with web services. In: Proceedings of iiWAS '14, 4–6 December 2014, Hanoi, Vietnam. doi:10.1145/2684200.2684310

31. Sarelo, K.: A SOA for ubiquitous communication management. In: Proceedings of iiWAS2009, 14–16 December 2009, Kuala Lumpur, Malaysia. doi:10.1145/1806338. 1806386

32. Baskin, A., et al.: Exploring the role of finiteness in the emergence of structure. In: Mittenthal, J., Baskin, A. (eds.) The principles of organization in organisms. Santa Fe Institute studies in the sciences of complexity, Proceedings vol 13. Addison-Wesley, Reading, pp. 337–377 (1992)

33. Li, M., Vitanyi, P.M.B.: Two decades of applied Kolmogorov complexity: in memoriam of Andrei Nikolaevich Kolmogorov 1903–1987. In: Proceedings of 3rd annual structure in complexity theory conference, Georgetown University, Washington, 14–17 June 1988

34. Mittenthal, J.E., et al.: Patterns of structure and their evolution in the organization of organisms: modules, matching, and compaction. In: Mittenthal, J., Baskin, A. (eds.) The principles of organization in organisms. Santa Fe Institute studies in the sciences of complexity, Proceedings vol. 13. Addison-Wesley, Reading, pp 321–332 (1992)

35. Kauffman, S.A.: The sciences of complexity and "origins of order". In: Mittenthal, J., Baskin, A. (eds.) The principles of organization in organisms. Santa Fe Institute studies in the sciences of complexity, Proceedings vol 13. Addison-Wesley, Reading, pp. 303–319 (1992)

36. Suh, N.P.: Axiomatic Design. Oxford University Press, New York (2001)

37. Lu, S.C.Y., et al.: A scientific foundation of collaborative engineering. CIRP Ann. Manufact. Technol. **56**(2), 605–634 (2007). doi:10.1016/j.cirp.2007.10.010
38. Chan, T-W., Baskin, A.: Studying with the prince: the computer as a learning companion. In Proceedings of the ITS-88 Conference (1988), pp. 194–200
39. Graesser, A.C., et al.: AutoTutor: an intelligent tutoring system with mixed-initiative dialogue. IEEE Trans. Educ. **48**(4), 612–618 (2005). doi:10.1109/TE.2005.856149

Chapter 6
Leveraging Analytics for Digital Transformation of Enterprise Services and Architectures

Alfred Zimmermann, Rainer Schmidt, Kurt Sandkuhl, Eman El-Sheikh, Dierk Jugel, Christian Schweda, Michael Möhring, Matthias Wißotzki and Birger Lantow

Abstract The digital transformation of our society changes the way we live, work, learn, communicate, and collaborate. The digitization of software-intensive products and services is enabled basically by four megatrends: Cloud Computing, Big Data Mobile Systems, and Social Technologies. This disruptive change interacts with all information processes and systems that are important business enablers for the current digital transformation. The Internet of Things, Social Collaboration Systems for Adaptive Case Management, Mobility Systems and Services for Big Data in Cloud Services environments are emerging to support intelligent user-centered and social community systems. Modern enterprises see themselves confronted with an ever growing design space to engineer business models of the future as well as their IT support, respectively. The decision analytics in this field becomes increasingly complex and decision support, particularly for the development and evolution of sustainable enterprise architectures (EA), is duly needed. With the advent of intelligent user-centered and social community systems, the challenging decision processes can be supported in more flexible and intuitive ways. Tapping into these systems and techniques, the engineers and managers of the enterprise architecture become part of a viable enterprise, i.e. a resilient and continuously evolving system that develops innovative business models.

A. Zimmermann (✉) · D. Jugel · C. Schweda
Reutlingen University, Reutlingen, Germany
e-mail: alfred.zimmermann@reutlingen-university.de

R. Schmidt
Munich University, Munich, Germany

K. Sandkuhl · D. Jugel · M. Wißotzki · B. Lantow
University of Rostock, Rostock, Germany

E. El-Sheikh
Center for Cybersecurity, University of West Florida, Pensacola, FL, USA
e-mail: eelsheikh@uwf.edu

M. Möhring
Munich University, Munich, Germany

© Springer International Publishing Switzerland 2016
E. El-Sheikh et al. (eds.), *Emerging Trends in the Evolution of Service-Oriented and Enterprise Architectures*, Intelligent Systems Reference Library 111,
DOI 10.1007/978-3-319-40564-3_6

6.1   Introduction

Information, data and knowledge are fundamental concepts of our everyday activities. Social networks, smart portable devices, and intelligent cars, represent only a few instances of a pervasive, information-driven vision [1] for the next wave of the digital economy and better-aligned information systems. Digitization [2] encompasses the collaboration of human beings and autonomous objects beyond their local context using digital technologies. Digitization further increases the importance of information, data and knowledge as fundamental concepts of our everyday activities. By exchanging information human beings and intelligent objects are able to make decisions in a broader context and with higher quality. Major trends for digital enterprise transformation are investigated by Leimeister et al. [3]: (i) Digitization of products and services: products and services are enriched with value-added services or are completely digitized; (ii) Context-sensitive value creation: though popularity of mobile devices location contexts are used more frequently and enable on demand customized solutions; (iii) Consumerization of IT: One of the challenges is the safe integration of mobile devices into a managed enterprise architecture for both business and IT; (iv) Digitization of work: Today it is much easier to work together over large distances, which allows often an uncomplicated outsourcing of business tasks; and the (v) Digitization of business models: Businesses need to adapt and have to rethink their business models to develop innovative business models according to employees' current skills and competencies.

The technological and business architectural impact of digitization has multiple aspects, which directly affect adaptable digital enterprise architectures and their supported systems. Smart companies are extending their capabilities continuously managing their changing Business Operating Model [4] by developing and maintaining Enterprise Architectures as the architectural part of a changing IT Governance [5]. Enterprise Architecture Management [6–8] and Services Computing [9, 10] is the approach of choice to organize, build, utilize, and distribute capabilities for digital enterprise architectures [11, 12]. They provide flexibility and agility in business and IT systems. The development of such applications integrates Web and REST Services, Cloud Computing and Big Data management, among other frameworks and methods for the architectural semantic support. Today's information systems span a broad range of domains including: intelligent mobility systems and services, intelligent energy support systems, smart personal health-care systems and services, intelligent transportation and logistics services, smart environmental systems and services, intelligent systems and software engineering. One of the challenges is the safe integration of mobile devices into managed enterprise architecture of both business and IT. Today it is much easier to work together over large distances, which allows often an uncomplicated outsourcing of business tasks. Businesses need to adapt and have to rethink their business models to develop innovative business models according to employees' current skills and competencies.

Digitization of products and services requires the close alignment of business models and digital technologies for creative digital strategies and solutions, as well as for their digital transformation. Unfortunately, the current state of art and practice of enterprise architecture lacks an integral understanding and support of collaborative decisions in the process of architectural adaptation and enterprise transformation. We have therefore to extend previous approaches of enterprise architecture to fit to the digitization of new products and services and by introducing suitable mechanisms for collaborative architectural engineering and decision support with adaptive case management for agile changing business models, information systems and their digital enterprise architecture.

We are investigating concepts and mechanisms for analyzing enterprise architectures to provide decision support for the architectural evolution and adaptation. We abstain from defining a heavyweight framework for EA management, but provide a platform laying a basis for manifold analysis techniques that can be combined as necessary. We regard this approach advantageous over the state-of-the-art with its abundance of "ingredients" that are often not adopted in the practice of EA management. Further, the analysis techniques allow a focus on key aspects of the ongoing transformation, like a cloud transformation, without losing the enterprise context, which is represented in different perspectives and stakeholder-specific viewpoints.

A new refocused decision-oriented approach for digital enterprise architectures should be both holistic and easily adaptable. Our aim is to support flexibility and agile transformations for both business domains and related enterprise systems through semiautomatic semantic-supported decisional processes, which are combined with analytics of real-time changing information environments. The present research is focused on decisional support for conceptual and architectural information, analytics-based methods, semantic representations and inference mechanisms, which are combined to enable stakeholder-centric decisional processes and transparency information for digital transformations.

Section 6.2 describes our fundamental orientation for digitized products and services. Section 6.3 focuses on our research platform for digital enterprise architecture, which was extended by concepts from adaptive case management, mechanisms for architectural adaptation and a specific model integration method. Section 6.4 presents our decision case management environment and links this in Sect. 6.5 with collaborative decision services and mechanisms. In Sect. 6.6 we present the decisional metamodel for digital enterprise architectures as a base for our decision analytics approach. Section 6.7 sketches the semantic support for architectural analytics by adding a suitable knowledge representation for both architectural concepts and decision metamodels. Finally, we summarize in Sect. 6.8 our findings and our future research plans.

6.2   Digitization of Products and Services

Digitized products and digitized services are both software-intensive and therefore malleable and usually service-oriented. They are able to increase their capabilities accessing cloud-services and change their behavior. Digitized products support the co-creation of value together with the customer and other stakeholders. Digitized products and services offer disruptive opportunities for new business solutions having new smart connected functionalities. At first, the high level of interest surprises, because the digital representation of information and performing digital calculation operations have been established for decades. The term digitization has its origin in [13] and is used for the digital representation of information, and processing since years [14].

There are definitions that consider digitization a primarily technical term [15]. Technologies often associated with digitization [16] are: cloud computing [17], big data [18, 19] advanced analytics, social software, and the Internet of Things [20]. The set of technologies increases. New technologies such as deep learning [21] are emerging that allow computing to be applied to activities that were considered as exclusive to human beings.

Therefore the question arises, what causes the present emphasis on digitization and what is different about digitization. Out thesis is, that digitization today embraces effects from both a product, and a value-creation perspective. Digitization can be described from both a product and a value-creation perspective: digitized products and the digitized value chains. Digitized products offer new capabilities to interact with their environment and the customer. They are also capable to collect data.

Classic industrial products are static [22]. You can change the production not or only to a limited extent. Digitization creates products containing software that can be upgraded via network connections. In addition, products over network connections can use external services. Software and especially services are also easier to update. New software functions can be added and additional services can be integrated. Therefore, the functionality of products is no longer static, but can be adapted to changing requirements and hidden customer needs. In particular, it is possible to create digitized products and services step-by-step or provide temporarily unlockable functionalities. So, customers whose requirements have risen can add functions without hardware modification.

Digitalization [2] allows products to capture their own state and submit this information into linked contexts. The provider can remotely determine whether the product is still functional and encourage, where appropriate, maintenance and repairs. This is the basis on which, instead of the physical product, the use of the product as a service changes the traditional offer. These services will be measured on their effectiveness and their practical usage. This will lay the foundation for usage-based billing models. In addition to the usage information also the condition of the product by the manufacturer can be queried.

In this context, concepts of preventive maintenance [23] can be developed. These have the objective of unscheduled stoppages whenever possible to avoid. Evaluation of status information and analysis of the history of use of the product can be predicted, when a malfunction of the product is likely. A maintenance or replacement of the product is performed before the respective date. In this context, the collected data can also be used to provide information for a repair on the spot, so that a high first time solution rate can be achieved. At the same time, storage can be improved in this way of spare parts.

The Internet of Things [24] enables the creation of products that are constantly in communication with the manufacturer. In this way, the manufacturer can win genuine information about the use of the product. The collection of information on the use of products is no longer dependent on the cooperation with the customer. In addition, it is possible to collect important information for up—and cross—selling in this way. By linking devices on networks, benefits are generated from two areas. Both the functionality increases and there are positive effects arising from the overarching data use. Furthermore, the production of more customer-oriented products [25] is possible.

Network effects [26] grow exponentially, because they are based on the number of participants and the number of possible connections. The possibility to connect devices of the network increases the possibilities of the individual device, because increasing the number of potential partners. This benefit increase is disproportionate higher as the number of devices, since the number of possible connections grows faster as the device number [27].

This increase of commercial value also happens through services provided by a lot of partners with complementary skills [22]. Software platforms that support the collection, analysis and exchange of data are rapidly growing. Winners in this environment will be companies, enable network effects to create value for customers. Network effects become apparent not only in functionality, but also in the scope of the data. These effects are called network intelligence [13]. By bringing together data from different network nodes, trends can be detected much earlier and more accurately.

By linking data from different sources [28], it is possible to establish correlations that would not have been possible with the data of a single device. This effect increases with the number of devices. By integrating external data sources, the extraction of relevant information can be improved also. Particularly the ability of big data and advanced analytics helps to process particularly semi- and unstructured data.

Characteristic is the involvement of individual product in an information system, which accelerates the learning and knowledge processes across all products [19]. In this way, a number of other beneficial effects can be achieved as network optimization, maintenance optimization, improved restore capabilities, and additional evidence against the consideration of individual systems.

Central is the idea that the producer of goods creates value and the value is determined at the moment of exchange of goods. It was tried to transfer this idea on services. However, this led to a service definition, which considers services as a negation of physical goods [29]. Services are not material, but already the missing homogeneity can be challenged for industrial services. Services are also not divisible, i.e. they must be provided as a whole. Services are also not durable; they are not stored and are provided only at the moment of need.

Basis for the implementation of the co-creation [30] approach of service-dominant logic is the continuous connection of the products with the manufacturer. The manufacturer can win genuine information about the use of the product. Important information for the development of new products can be obtained in this way. The consumer converts dynamically to be co-producer [31]. Platforms are complementary products, which cooperate via standardized interfaces [32]. Since the development of new functionalities by different partners is distributed [33] platforms significantly speed-up the development time of new solutions.

6.3 Digital Enterprise Architecture

Enterprise Architecture Management (EAM) [6, 7, 9, 34] defines today with frameworks [35], languages [36], and standards [37, 38], tools and practical expertise a quite large set of different views and perspectives. EAM can be e.g. used to support and implement business processes as well as to reach business goals [39]. Benefits of EAM are influenced by different influence factors such as EAM knowledge, landscape complexity and Business IT alignment [40]. We argue in this paper that a new refocused digital enterprise architecture approach should support digitization of products and services, and should be both holistic [41, 42] and easily adaptable [43] to support the digital transformation with new business models and technologies like social software, big data, services and cloud computing, mobility platforms and systems, security systems, and semantics support. We are extending the first versions of ESARC–Enterprise Services Architecture Reference Cube [41, 42] (Fig. 6.1).

In this paper we extend our service-oriented enterprise architecture reference model for the context of managed adaptive cases and decisions [44, 45], which are supported by case services of a collaborative case framework [44] within an adaptive case management environment [46]. Additionally we have considerably extended our architectural metamodel integration approach [47] to support digital enterprise architectures for digital transformations [11] and the integration of Internet of Things [12, 24] architectures.

ESARC—Enterprise Services Architecture Reference Cube [41, 42] is an architectural reference model for an extended view on evolved digital enterprise architectures. ESARC is more specific than existing architectural standards of EAM—Enterprise Architecture Management [35, 36] and extends these architectural standards for digital enterprise architectures with services and cloud computing. ESARC provides a holistic classification model with eight integral architectural domains. These architectural domains cover specific architectural viewpoint descriptions [37, 38] in accordance to orthogonal dimensions of both architectural

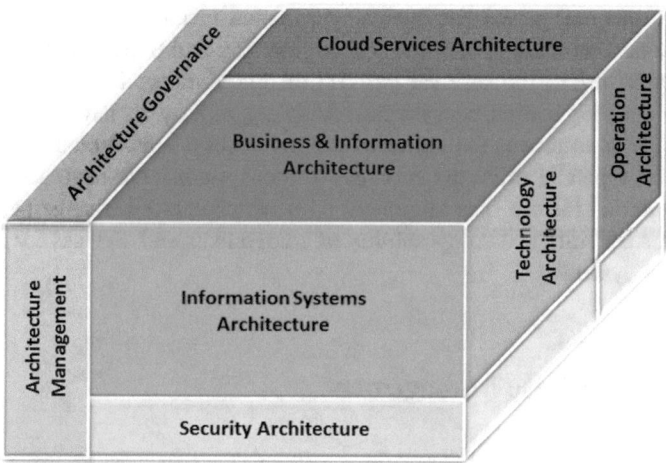

Fig. 6.1 Enterprise services architecture reference cube [41–43]

layers and architectural aspects [6, 34, 42]. ESARC abstracts from a concrete business scenario or technologies, but it is applicable for concrete architectural instantiations to support digital transformations. The Open Group Architecture Framework [35] provides the basic blueprint and structure for our extended service-oriented enterprise architecture domains of ESARC [41, 43] having: Architecture Governance, Architecture Management, Business and Information Architecture, Information Systems Architecture, Technology Architecture, Operation Architecture, and Cloud Services Architecture. ESARC provides a coherent aid for examination, comparison, classification, quality evaluation and optimization.

We developed an architectural evolution approach to integrate and adapt most valuable parts of existing EA frameworks and metamodels from theory and practice [47]. Additionally to handling architectural structures for dynamically extending core metamodels we see a chance to integrate decentralized mini-metamodels, models and data of architectural descriptions coming from small devices and new decentralized architectural elements, which traditionally are not covert by enterprise architecture environments. The focused model integration approach is based on special correlation matrixes to identify similarities between analyzed model elements from different provenience and integrate them according their most valuable contribution for an integrated model. According to [48] we are building the conceptualization of EA in 4 steps—from stakeholders' needs, to the concerns of stakeholders, then the extraction of stakeholder relevant concepts, and last but not least the definition of relationships for new tailored architectural metamodels.

Our research consists of a metamodel-based model extraction and integration approach [47] for digital enterprise architecture viewpoints, models, standards, frameworks and tools to support digital transformations [11, 12]. Currently we are working on the idea of continuously integrating small EA descriptions for relevant objects of digital enterprise architecture. These EA-Mini-Descriptions consists of partial EA data and partial EA models and related metamodels. Our goal is to be able to support an integral architectural engineering and transformation process.

Adaptation drives the survival [49–51] of digital enterprise architectures [47], platforms and application ecosystems. Adapting rapidly to new technology and market contexts improves the fitness of adaptive ecosystems. Volatile technologies and markets typically drive the evolution of ecosystems. We have additionally to consider internal factors. The alignment of Architecture-Governance [4, 5] shapes resiliency, scalability and composability of components and services for distributed information systems.

6.4   Decision Case Management

A Decision support system (DSS) is a system "[…] to help improve the effectiveness of managerial decision making in semi-structured tasks" [52, p. 255], and according to [53]. In particular knowledge intensive management activities, like EAM, can benefit from a DSS to improve architectural decision-making. In the following we explore how an EA cockpit [54] can be leveraged and extended to a DSS for EAM. A cockpit presents a facility or device via which multiple viewpoints on the system under consideration can be consulted simultaneously. Each stakeholder who takes place in a cockpit meeting can utilize a viewpoint that displays the relevant information. Thereby, the stakeholders can leverage views that fit the particular role like Application Architect, Business Process Owner or Infrastructure Architect [55]. The viewpoints applied simultaneously are linked to each other such that the impact of a change performed in one view can be visualized in other views as well. Figure 6.2 gives the idea of an example architectural cockpit.

Jugel et al. [56] present a collaborative approach for decision-making for EA management. They identify decision making in such complex environment as a knowledge-intensive process strongly depending on the participating stakeholders. Therefore, the collaborative approach presented is built based on the methods and techniques of adaptive case management (ACM).

Adaptive Case Management (ACM) [44, 45] offers a lightweight model to support knowledge-intensive processes, which are driven by user decision-making.

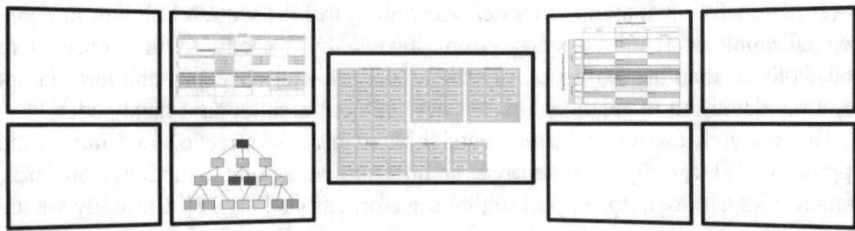

Fig. 6.2 Example: enterprise architecture cockpit [54, 56]

Knowledge processes of usually high-skilled stakeholders, like enterprise architects, require process adaptations at run-time. ACM is not dictating a predefined course of action [57] and provides the necessary information and knowledge support to be able to solve a case. A case [45] is typically a collection of all relevant information into one place, which is handled by one or more knowledge workers during solving this case. The case is the jointly used focal point for assessing the situation, initiating activities and processes, implementing the work, and reflecting results based on a history record about what was really done. A case brings together all the necessary resources and also tracks everything that has happened into a record history, which can be mined to synthesize best practices, patterns of success, and used and extended instruments. Fundamental aspects and requirements for ACM, are mentioned in [57]:

1. The adaptation aspect of ACM consists of content, people, and reporting capabilities to be able to change the knowledge process at run-time by end-users. Additionally to the adaptation aspect a knowledge worker should be able to continuously improve his case templates.
2. The organization aspect groups policies, processes, and data. In ACM data is the dominant factor as opposed to the process-oriented view from BPM. Knowledge work requires the integration of data [50] into the execution process.
3. The case handling aspect is about collaboration, decision support, and integration of resources, events, and communication. Complex problems are typically solved collaboratively by involving individual stakeholders in respect of different necessary knowledge types and stakeholder concerns. Decision support requires transparency within a shared understanding of analyzed EA scenarios by named stakeholders.

Opposed to routine work, which can be supported by business process management because of its repeatable kind, knowledge work is typically unpredictable. Knowledge workers [58, 59] are acting under uncertainty. An unpredictable process [45] does not repeat in routine patterns and emerges as the work is done. The practice of preparing for many possible courses is called agility. Differentiating seven domains of predictability [45] case management can be focused on two main types:

1. Product Case Management: Supports design-time knowledge processes with a well-known set of actions, having much variation between individual cases. It is not possible to set out a single fixed process. Knowledge workers are actively involved in deciding the course of events for a case.
2. Adaptive Case Management: Knowledge workers are involved not only in the case, and picking predefined actions, but they are constantly adapting the process and striving for innovative approaches, and may want to share and discuss process plans.

The Case Management Modeling Notation (CMMN) [60] is a notation for ACM that describes mandatory and optional tasks (DiscretionaryItem), and thereby

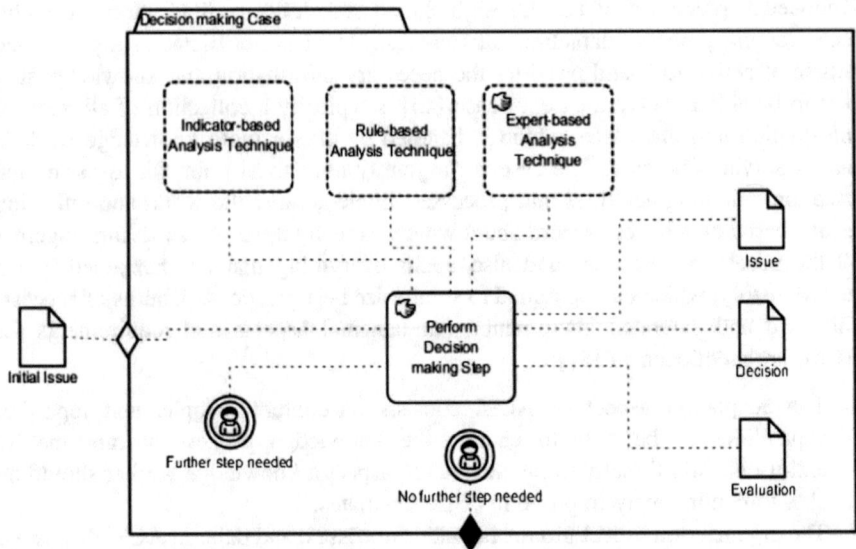

Fig. 6.3 CMMN model of collaborative decision making case [61]

supports flexible processes. In line with Jugel et al. [61], we utilize the CMMN to describe a collaborative decision-making case for EAM, cf. Fig. 6.3.

The *Issue* is the starting point of a collaborative decision-making case. This issue describes the problem space of the decision-making activity, which aligns with the perspective of Mayring [62]. We further assume that goals and success criterions, as required by Johnson and Ekstedt [63], have already been defined as part of strategic management activities. The issue is the reason why the EA has to be analyzed and decided upon. Based on this issue, involved stakeholders choose viewpoints that they need to analyze the issue.

The *decision-making step* is the central activity of the decision-making case presented in Fig. 6.3. This step can involve different optional activities in which different kinds of quantitative and qualitative analysis techniques [64] are applied to gain additional insights [60]:

- *Expert-based analysis* techniques are dependent on expert knowledge and tacit information of the involved stakeholders. Jugel and Schweda [54] identify these techniques with interactive functions like "graphical highlighting and filtering".
- *Rule-based analysis* techniques correspond to algorithms that are used to indentify patterns in the EA. Hanschke provides so-called analysis patterns in [65], which are examples of rule-based analysis techniques.
- *Indicator-based analysis* techniques are formal methods that compute indicators from properties of the EA. Matthes [66] present quantitative, metrics-driven EA analyses by quantitatively assessing architectural properties and therefore use an indicator-based analysis technique.

The stakeholders apply different of these techniques in the decision-making step and interpret the results of the techniques for additional insights [62]. While performing a decision-making step, stakeholders can choose analysis techniques, which are part of a catalog. The catalog is independent of a particular case. After choosing an analysis technique, it is performed. In case of rule-based and indicator-based analysis techniques, the techniques can be performed automatically using algorithms and aggregations. In case of an expert-based analysis technique, stakeholders must manually analyze the EA by using and interaction with the cockpit's views.

The *decision-making step* is based on case data consisting of an EA model and additional insights elicited in previous steps. Consequently, the insights gained during each step contribute to the *case file* (CaseFile) of the decision-making case. Derived values, like the values of KPIs are thereby not considered additional information, but only a different way of representing and aggregating existing information. If stakeholders based on the values of a KPI decide on affected architecture elements, these decisions and considerations represent new information, which is added to the case file. In particular, the stakeholders' interpretation can yield following additional elements for the case file (CaseFileItem):

- An evaluation represents the stakeholder's opinion on the analysis results.
- A new issue refines the previously analyzed one based on the analysis.
- A decision reflects a design alternative that is useful to resolve the issue.

During the decision-making, alternative designs can be identified [63]. In the final step of the decision-making process, not all previously evaluated designs will prevail. At the end of every *decision-making step*, the stakeholders have to decide, whether additional information is required or not—represented by to UserEventListeners in the CMMN diagram in Fig. 6.3. The case file of the decision-making case has to be structured appropriately to accommodate for the decision-making process.

The Object Management Group (OMG) has published the Case Management Model and Notation (CMMN) [60] as a first step to support modeling for case management scenariosmanagement scenarios. A case study of a TOGAF-style process [35] for EAM with CMMN was implemented in [67]. The upcoming standard Decision Model and Notation (DMN) of OMG [68] discern three usage models: for modeling human decision-making, for modeling requirements for automated decision-making, and for implementing automated decision-making. DMN bridges the gap between business decision designs and their implementation by providing a common notation for decision models. The purpose of DMN is to facilitate a decision model framework, which is easily usable for decision diagrams and as a base for optionally automating decisions. Decision-making support is addressed from basically two perspectives: normal BPMN business Process Models can be expanded by defining specific decision tasks, or decision logic can be used to support individual decisions, e.g. business rules, decision tables, or executable analytic models. DMN can additionally provide a third perspective to bridge between business process models and decision logic by introducing the Decision Requirements Diagram. Complementary to the DMN notation, which is used to model decisional relationships and concepts like Decision, Input Data, Business Logic, Application, Application Risk, etc. DMN introduces an expression language to represent decision tables, decision rules, and function invocations. Today we are exploring the suitable usage and close link of DNM for decisional support logic within our architectural engineering and analytics research.

6.5 Collaborative Decision Processes

Although concepts such as Business Process Management [69] introduced a customer-oriented perspective, it still contains many concepts following the ideas developed already in [70]. These are the division of larger tasks into defined, smaller tasks and the assignment of individual responsible to accomplish these tasks. Therefore it does not surprise, that a plenty of approaches such as [71], Swenson [44] tried to develop support for cooperation beyond strictly structured business processes as almost all WFMSs and most of the BPMSs, but also some groupware and case management systems. However these approaches become not as successful as expected.

One has to meet a number of challenges when supporting EA management processes. The first challenge is the lack of a pre-defined workflow. Similar to adaptive case management [44, 45] the control-flow of EA management processes cannot be predefined in most situation. Instead the control-flow is defined "on-the-fly" during execution of the EA management process.

The second challenge is organizational integration [72]. Many early approaches addressing the support of EA management processes limited the participation of stakeholders. E.g. although classical groupware abstained from pre-defining a strict control flow, specific access rights to documents had been assigned. Thus the group of possible contributors had been limited. In this way an a priori-decision had been made deciding who may contribute and who may not. Some stakeholders were not able to contribute.

The third challenge is semantic integration [72]. Due to the involvement of a multitude of stakeholders, semantic frictions such as homonyms and synonyms create misunderstandings between the process participants. These semantic frictions may delay the EA management process or even worse, may cause deficient architectures.

Social software is based on four basic principles: social production [73], weak ties [74], collective decisions [75], and value co-creation [76]. Each of these principles support EA management processes by addressing one or more challenges, as addressed in Fig. 6.4.

Social production [73] is the creation of artifacts without a top-down created plan but by combining the suggestions and decisions from independent contributors. By abstaining from Tayloristic top-down planning, new and innovative

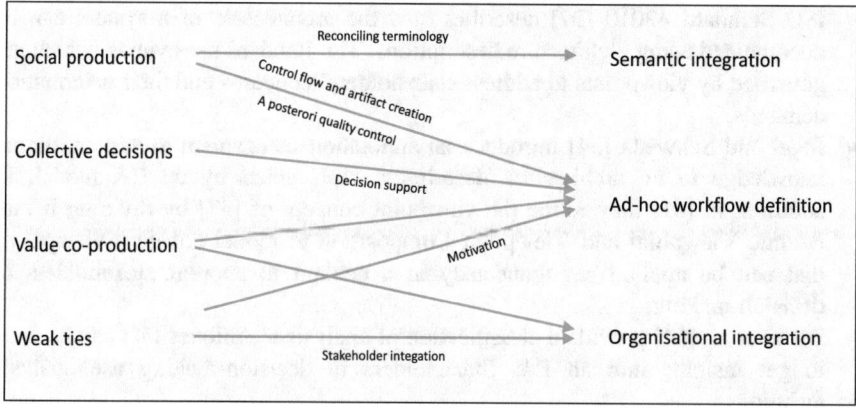

Fig. 6.4 Collaborative engineering and transformation [77]

contributions outside the original scope can be identified and added. Due to these properties, social production matches the requirements of EA management processes. The control flow of EA management processes can be defined in an ad hoc manner. During execution of the EA process, architectural artifacts can be investigated in a cooperative way.

Collective decisions [75] provide a new way in EA management processes to make decisions. They provide statistically better results than experts, if the decision cannot be made using scientific means and the participants decide independently. Surowiecki describes in [78] the approach of the so-called the wisdom of crowds. He argues that a decision made by several persons often leads to better results, because each person has a specific knowledge. Value-co-production [78] is also supporting the definition and execution of EA management processes by integrating contributions from the business side. By abolishing the separation between artifact producer and consumer, a better adaptation to the individual requirements can be achieved. Furthermore value co-production enhances the organizational integration.

6.6   Decision Analytics

In this section we present a decisional metamodel based on the work of Jugel et al. [61] to support the decision-making case presented in the previous section. The metamodel focuses on the documentation of decision and rationalizing information and is a combination of several approaches that partly cover aspects of decision-making.

- Plataniotis et al. [79] describe an approach called "EA Anamnesis" focusing on ex-post modeling EA decisions and decision-making strategies. However, they do not describe decision processes. Furthermore, they do not describe rationales.

- ISO Standard 42010 [37] describes how the architecture of a system can be documented using architecture descriptions. The standard uses views, which are governed by viewpoints to address stakeholders' concerns and their information demands.
- Jugel and Schweda [54] introduce an annotation mechanism to add additional knowledge to an architecture description represented by an EA model. In addition, in [61] they refine the viewpoint concept of [37] by dividing it into Atomic Viewpoint and Viewpoint Composition to model coherent viewpoints that can be applied simultaneously in a cockpit to support stakeholders in decision-making.
- Buckl et al. [64] provide a classification of analysis techniques that can be used to get insights into an EA. Stakeholders in decision-making use analysis techniques.

The Case Management Modeling Notation (CMMN) [60] is a notation for ACM to describe flexible processes including optional tasks. The notation provides us base concepts to model cases.

Figure 6.5 illustrates the decisional metamodel. The background colors of the concepts indicate their origin. Green colored concepts have their origin in ISO Standard 42010 [37], gray colored concepts in "EA Anamnesis" [79], blue colored concepts in CMMN [60], yellow colored concepts in [64] and red colored concepts in [54, 61]. The decisional metamodel focuses on the stakeholders using viewpoints to perform a Decision making Step that is in line with CMMN [60] a HumanTask. During this step, stakeholders have the ability to choose Analysis Techniques that are in line of CMMN [60] DiscretionaryItems. Additional information during a step is created and persisted as Annotations to the deci-sional views. Annotations as well as Views are in line with CMMN [60] CaseFileItems, because both represent relevant information within a case and are therefore part of a CaseFile. The annotation concept aligns with the one presented by Jugel et al. in [54] and reflects different EA issues (also the initial one of the decision case), Evaluations of the

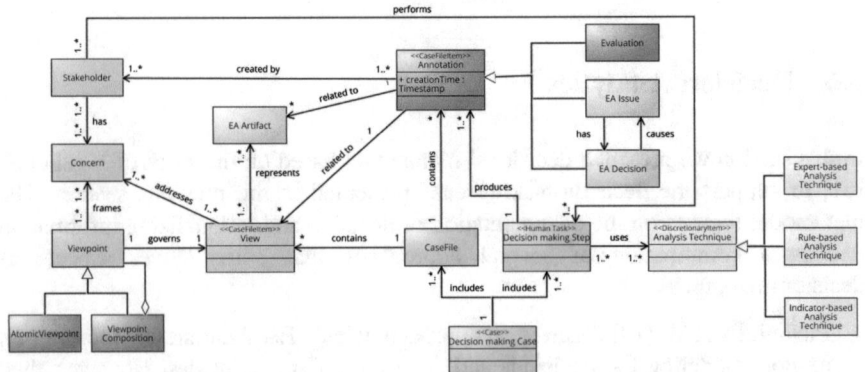

Fig. 6.5 Collaborative EA decision making metamodel

analyses' results, and EA decisions. As the annotations can be based on the results of an analysis technique, also the applied techniques are part of the metamodel and are persisted in the CaseFile. Latter notion corresponds to the terminology of CMMN.

For the utilized viewpoints, we distinguish between Atomic Viewpoints and Viewpoint Compositions [61]. Whereas an Atomic Viewpoint is a single Viewpoint in line with ISO Standard 42010 [37], a Viewpoint Composition forms a composite structure and consists of coherent Atomic Viewpoints or other Viewpoint Compositions needed by Stakeholders to satisfy their information demands elicited by their Concerns. Viewpoint Compositions are assembled to address a specific decision-making case from multiple perspectives. A cockpit, as presented in the previous chapter, is a viewpoint composition.

In addition, Annotations are the triggers for the next Decision Making Step. One or more Stakeholders are responsible for a step and perform them. Within Decision Making Step stakeholders can choose between different Analysis Techniques to get additional information needed to satisfy their information demands. Analysis Techniques are based on Annotations as well on the EA model. Annotations describe additional information related to EA Artifacts. EA Issues and EA Decisions, as already present in the model of Plataniotis et al. [79], represent additional knowledge and are therefore specializations of Annotation. As described in [79], EA Decisions can be decomposed, translated and substituted into other EA Decisions. Modeling alternatives is also possible. According to our decision-making case, we added Evaluation as a third sub-concept of Annotation.

6.7   Semantic Support for Architectural Analytics

Semantic technologies [80] in general and ontologies in particular provide support for architectural analytics in many ways. The general features of ontologies address some of the aforementioned challenges and requirements for architectural analytics. Namely, the provision of domain specific knowledge and vocabulary allows the creation of stakeholder specific views (cf. Sect. 6.6), ontology alignment and mapping are common mechanisms of semantic integration (cf. Sect. 6.5), inference on ontologies can identify patterns in the domain knowledge (cf. Sect. 6.4) and can make implicit knowledge explicit by adding new facts to the knowledgebase.

Thus, many approaches have been made to represent enterprise models or enterprise architecture respectively by creating ontologies for this domain. The most popular examples are probably Uschold et al.'s "The Enterprise Ontology" [80] and Dietz's DEMO approach [8]. Further publications in the area are [81–83]. The Enterprise Architecture Ontology for Services Computing from [83] extends the ESARC metamodel from [41] with e semantic representation for enterprise architectures. Ontologies in enterprise modeling and architecture are useful, as shown by Sandkuhl et al. [84].

The most widely used definition of ontologies in computer science characterizes ontologies as "formal, explicit specification of a shared conceptualization" [85]. Here, "conceptualization" means creating an abstract model of real world phenomena by identifying relevant concepts of them. "Explicit" refers to a clear definition of concepts, concept types, and the constraints on their use. "Formal" means that an ontology is machine readable, and "shared" reflects the intention that the ontology should be a consensus, accepted within the communities.

Literature defines several functionalities and features of ontologies that support enterprise architecture analytics. Uschold and Gruninger [86] names three uses of ontologies: (1) Communication, (2) Interoperability, and (3) Systems engineering. Since ontologies are shared conceptualizations in communities they provide a base for human communication. Being a normative, assuring consistency, and avoiding ambiguity they foster knowledge exchange in collaborative decision scenarios, as they are present in enterprise architecture analytics scenarios. Furthermore, they provide networks of relationships that relate the knowledge regarding different concerns of stakeholders and they provide a semantic integration of this knowledge on the level of human communication. Interoperability or integration by ontologies provides the same features on the level of externalized knowledge in information systems. Thus a better information quality can be achieved for decision situations. At last, the use of ontologies for systems engineering assures reuse of existing knowledge and better interoperability of information systems. Bürger and Simperl additionally name in [87] specifically (4) Computational Inference and (5) Knowledge Reuse and Organization as contributions of ontology use. Computational inference allows for deriving implicit facts and logical inconsistencies. Having concepts and rules systematically formalized, reuse of models and model-party in different domains becomes possible.

Antunes et al. describe in [88, 89] specifically the use of ontologies in enterprise architecture analysis:

- Improved extensibility and expressiveness of the enterprise architecture through ontology based integration of domain-specific meta-models.
- Improved enforcement of meta-model coherence by defining constraints of concept use.
- Improved meta-model conformance verification by the use of reasoners that can identify logical inconsistencies in enterprise architecture models.
- Improved analysis for decision making through the use of inference and query mechanisms.

These features can be derived from the general features of ontologies. Subsuming the discussion, the benefits of ontologies for enterprise architecture analytics are twofold. First, they provide means for a better communication in the collaborative decision scenarios of enterprise architecture analytics. Second, they support rule-based analysis techniques by computational inference and a potentially broad information base through interoperability.

Antunes et al. propose in [81] a general four-step process of enterprise architecture analysis using ontologies:

1. Identify stakeholders and analysis needs: After identifying the stakeholders, their information needs are gathered in the form of questions and the expected type of answer (e.g. a list of processes or actors). Afterwards, the analysis of questions identifies relevant concepts and instances.
2. Review enterprise architecture models: A comparison between concepts needed for analysis and those in the enterprise architecture model is performed. If there is a gap, hence the model does no cover the stakeholders needs, new concepts have to be added by ontology engineering or by integration of domain specific ontologies.
3. Instantiate model: A model instance for a specific scenario is created.
4. Perform analysis: Computational inference mechanisms are used to answer the questions. In the concrete approach by Antunes et al. Description Logic (DL) queries are used. However, depending on the used tools other ontology query languages can be used.

Antunes et al. provide in [89] an investigation regarding the possibilities of supporting analysis types [4, 5] using DL. Reasoning tasks of DL are:

- Subsumption: Organizing concepts in taxonomy. Hence, finding the most specific super class for a given class.
- Instance checking: Verifying if an instance is a member of a specific class or represents a specific concept respectively.
- Relation checking: Verifying whether two instances are related to each other in a certain way.
- Concept consistency: Verifying that there is no contradiction in concept definitions or concept definition chains.
- Knowledgebase consistency: Verifying that there is no contradiction in the model instance.

Taking general ontology engineering approaches, such as Ontology 101 by Noy and McGuiness [90] into account, step 2 also includes the integration or definition of semantic rules that allow deriving implicit facts within the model instance. Thus, queries performed in step 4 may also refer to facts that have been added to the model instance by computational inference. Antunes et al. show the practical applicability of their approach in [89] by the analysis of an ArchiMate [36] model using the Domain Independent Ontology (DIO, representing a conceptualization of the ArchiMate [36] meta-model) and an integration of Domains Specific Ontologies (DSO, representing concepts used by specific stakeholders).

Besides these general steps for ontology based enterprise architecture analysis, it remains unclear where the use of this approach is appropriate and where not. Two dimensions have to be considered answering this question. First, a classification of possible analysis tasks is needed. Different approaches can be used here; analysis patterns by Hanschke [65], and analysis dimensions by Buckl et al. [64] are prominent examples. Second, a classification of reasoning tasks that can be performed in ontologies is needed. Assuming commonly used OWL ontologies, the reasoning tasks supported by Description Logic (DL) cover the reasoning potential.

6.8 Conclusions and Future Work

In this paper we have identified the need for an integral understanding and support of collaborative decisions in the process of architectural adaptation and enterprise transformation. According to our research approach we have leveraged a new model of extended digital enterprise architecture, which is well suited for adaptive models and transformation mechanisms. We have extended the previous more static defined basic enterprise reference architecture by new metamodel elements for supporting cooperative decisions using mechanisms from adaptive case management. Related to our second research question we have presented our approach for collaborative processes in architectural engineering and transformation endeavors. We have combined architectural engineering and transformation processes with elements from adaptive case management. We have adapted typical architectural engineering processes with elements from social production, collective decision-making, value co-production, and week ties. Adaptive case management offers a lightweight model for knowledge-intensive processes. We have merged them with user decision-making processes within cooperative distributed environments for enterprise architecture management. We have introduced suitable individual decision support models and embedded them into cooperative analysis and engineering environments.

We are currently working on extended decision support mechanisms for an architectural cockpit for digital enterprise architectures and related engineering processes. Future work will extend both mechanisms for adaptation and flexible integration of digital enterprise architectures as well as will extend decisional processes by extensions of rationales and explanations.

References

1. Aier, S., et al.: Towards a more integrated EA planning: linking transformation planning with evolutionary change. In: Proceedings of EMISA 2011, pp. 23–36, Hamburg, Germany (2011)
2. Schmidt, R., et al.: Digitization—A multi-perspective definition. In: Proceedings of IDEA 2015, ESOCC Taormina, Italy. Springer, Berlin (2015)
3. Leimeister, J.M., et al.: Research program "Digital Business Transformation HSG". In: Working Paper Services of University of St. Gallen—Institute of Information Management, No. 1, St. Gallen, Switzerland (2014)
4. Ross, J.W., et al.: Enterprise Architecture as Strategy—Creating a Foundation for Business Execution. Harvard Business School Press (2006)

5. Weill, P., Ross, J.W.: It Governance: How Top Performers Manage It Decision Rights for Superior Results. Harvard Business School Press (2004)
6. Lankhorst, M., et al.: Enterprise Architecture at Work: Modeling, Communication and Analysis. Springer, Berlin (2013)
7. Johnson, P., et al.: IT Management with Enterprise Architecture. KTH, Stockholm (2014)
8. Bente, S., et al.: Collaborative Enterprise Architecture. Morgan Kaufmann, Los Altos (2012)
9. Zhang, L.J., et al.: Services Computing. Springer, Berlin (2007)
10. Papazoglou, M.P.: Web Services & SOA. Pearson (2012)
11. Zimmermann, A., et al.: Evolving enterprise architectures for digital transformations. In: Zimmermann, A., Rossmann, A. (eds.) Lecture Notes in Informatics, vol. P-244, pp. 183–194, DEC 15, 25–26 June 2015, Böblingen, Germany (2015)
12. Zimmermann, A., et al.: Digital Enterprise Architecture—Transformation for the Internet of Things. EDOCW 2015 with SoEA4EE, 21–25 Sept 2015, Adelaide, Australia (2015)
13. Tapscott, D.: The Digital Economy: Promise and Peril in the Age of Networked Intelligence, vol. 1. McGraw-Hill, New York (1996)
14. Brynjolfsson, E.: Understanding the Digital Economy: Data, Tools, and Research. The MIT Press, Cambridge (2000)
15. Weill, P., Woerner, S.: Thriving in an increasingly digital ecosystem. MIT Sloan Management Review, June 2015
16. Westerman, G., Bonnet, D.: Revamping your business through digital transformation. MIT Sloan Management Review, Feb 2015
17. Mell, P., Grance, T.: The NIST Definition of Cloud Computing. NIST (2011)
18. Agrawal, D., Das, S., El Abbadi, A.: Big data and cloud computing: current state and future opportunities. In: Proceedings of the 14th International Conference on Extending Database Technology, pp. 530–533 (2011)
19. Evans, P.C., Annunziata, M.: Industrial Internet: Pushing the Boundaries of Minds and Machines, General Electric, p. 21 (2012)
20. Atzori, L., Iera, A., Morabito, G.: The internet of things: a survey. Comput. Netw. **54**(15), 2787–2805 (2010)
21. Schmidhuber, J.: Deep learning in neural networks: an overview. Neural Netw **61**, 85–117 (2015)
22. Brynjolfsson, E., McAfee, A.: The Second Machine Age: Work, Progress, and Prosperity in a Time of Brilliant Technologies. W.W. Norton & Company (2014)
23. Duffuaa, S.O., Raouf, A.: Preventive maintenance, concepts, modeling, and analysis. In: Planning and Control of Maintenance Systems. Springer International Publishing, pp. 57–94 (2015)
24. Uckelmann, D., et al.: Architecting the Internet of Things. Springer, Berlin (2011)
25. Schmidt, R., et al.: Industry 4.0—Potentials for Creating Smart Products: Empirical Research results. In Abramowicz, W. (ed.) 18th International Conference on Business Information Systems, LNBIP 2008. Springer, Berlin, pp. 16–27 (2015)
26. Weitzel, T., et al.: Reconsidering network effect theory. ECIS 2000 Proceedings, Paper 91 (2000)
27. Metcalfe, B.: Invention is a flower, innovation is a weed. Tech. Rev. **102**(6), 54–57 (1999)
28. Provost, F., Fawcett, T.: Data Science for Business: What You Need to Know about Data Mining and Data-analytic Thinking, 1 edn. O'Reilly Media, Sebastopol (2013)
29. Vargo, S.L., Lusch, R.F.: The four service marketing myths: remnants of a goods-based manufacturing model. J. Serv. Res. **6**(4), 324–335 (2004)
30. Vargo, S., Lusch, R.: Service-dominant logic: continuing the evolution. J. Acad. Mark. Sci. **36**(1), 1–10 (2008)
31. Ritzer, G., Jurgenson, N.: Production, consumption, prosumption the nature of capitalism in the age of the digital 'prosumer'. J. Consum. Culture **10**(1), 13–36 (2010)
32. Baldwin, C.Y., Woodard, C.J.: The architecture of platforms: a unified view. In: Platforms, Markets and Innovation, pp. 19–44 (2009)

33. Eisenmann, T.R.: Managing proprietary and shared platforms. Calif. Manag. Rev. **50**(4), 31–53 (2008)
34. Iacob, M.-E., et al.: Delivering Business Outcome with TOGAF® and ArchiMate®. eBook BiZZdesign (2015)
35. The Open Group: TOGAF Version 9.1. Van Haren Publishing (2011)
36. The Open Group: ArchiMate 2.0 Specification. Van Haren Publishing (2012)
37. ISO/IEC/IEEE: Systems and Software Engineering—Architecture Description. Technical Standard (2011)
38. Emery, D., Hilliard, R.: Every Architecture Description needs a Framework: Expressing Architecture Frameworks Using ISO/IEC 42010. IEEE/IFIP WICSA/ECSA 2009, pp. 31–39 (2009)
39. Jonkers, H., et al.: Enterprise architecture: management tool and blueprint for the organization. Inf. Syst. Front. **8**(2), 63–66 (2006)
40. Schmidt, R., et al.: Benefits of enterprise architecture management—Insights from European experts. In: Proceedings of PoEM 2015, Valencia. Springer, Berlin (2015)
41. Zimmermann, A., et al.: Capability diagnostics of enterprise service architectures using a dedicated software architecture reference model. In: IEEE International Conference on Services Computing (SCC), pp. 592–599, Washington DC, USA, 2011
42. Zimmermann, A., et al.: Towards Service-oriented Enterprise Architectures for Big Data Applications in the Cloud. EDOC 2013 with SoEA4EE, pp. 130–135, 9–13 Sept 2013, Vancouver, BC, Canada (2013)
43. Zimmermann, A., et al.: Adaptable enterprise architectures for software evolution of SmartLife ecosystems. In: Proceedings of the 18th IEEE International Enterprise Distributed Object Computing Conference Workshops (EDOCW 2014), pp. 316–323, Ulm, Germany (2014)
44. Swenson, K.D.: Mastering the Unpredictable: How Adaptive Case Management will Revolutionize the Way that Knowledge Workers Get Things Done. Meghan-Kiffer Press (2010)
45. Swenson, K.D.: State of the Art In Case Management. White Paper Fujitsu (2013)
46. Collenbusch, D., et al.: Experiencing adaptive case management capabilities with cognoscenti. In: Zimmermann, A., Rossmann, A. (eds.) Lecture Notes in Informatics, vol. P-244, pp. 233–243, DEC 15, 25–26 June 2015, Böblingen, Germany (2015)
47. Zimmermann, A., et al.: Towards an integrated service-oriented reference enterprise architecture. In: ESEC/WEA 2013 on Software Ecosystem Architectures, pp. 26–30, St. Petersburg, Russia (2013)
48. Buckl, S., et al.: Modeling the supply and demand of architectural information on enterprise level. In: 15th IEEE International EDOC Conference 2011, pp. 44–51, Helsinki, Finland (2011)
49. Tiwana, A.: Platform Ecosystems: Aligning Architecture, Governance, and Strategy. Morgan Kaufmann, Los Altos (2013)
50. Heistacher, T., et al.: Pervasive Service Architecture for a Digital Business Ecosystem. arXiv preprint cs/0408047 (2004)
51. Bertossi, L.: Database Repairing and Consistent Query Answering. Morgan & Claypool Publishers (2011)
52. Keen, P.G.W.: Decision support systems: the next decade. In: Decision Support Systems, vol. 3(3), pp. 253–265. Elsevier, Amsterdam (1987)
53. Keen, P.G.W., Morton, M.S.S.: Decision Support Systems: An Organizational Perspective. Addison-Wesley, Reading (1978)
54. Jugel, D., Schweda, C.M.: Interactive functions of a Cockpit for Enterprise Architecture Planning. In: International Enterprise Distributed Object Computing Conference Workshops and Demonstrations (EDOCW 2014), pp. 33–40, Ulm, Germany (2014)
55. Wißotzki, M., Köpp, C., Stelzer, P.: Rollenkonzepte im Enterprise Architecture Management. In: Zimmermann, A., Rossmann, A. (eds.) Lecture Notes in Informatics, vol. P-244, pp. 127–138, DEC 15, 25–26 June 2015, Böblingen, Germany (2015)

56. Jugel, D., Kehrer, S., Schweda, C. M.: Providing EA decision support for stakeholders by automated analysis. In: Zimmermann, A., Rossmann, A. (eds.) Lecture Notes in Informatics, vol. P-244, pp. 151–162, DEC 15, 25–26 June 2015, Böblingen, Germany (2015)
57. Hauder, M., Pigat, S., Matthes, F.: Research challenges in adaptive case management: a literature review. In: International Enterprise Distributed Object Conference Workshops and Demonstrations (EDOCW 2014), pp. 98–107, Ulm, Germany (2014)
58. Fischer, L.: Taming the Unpredictable Real World Adaptive Case Management: Case Studies and Practical Guidance, Future Strategies (2011)
59. Fischer, L.: Empowering Knowledge Workers, Future Strategies (2014)
60. Object Management Group: Case Management Modeling Notation 1.0 (2014)
61. Jugel, D., Kehrer, S., Schweda, C.M., Zimmermann, A.: A decision-making case for collaborative enterprise architecture engineering. In: Cunningham, D., Hofstedt, P., Meer, K., Schmitt, I. (eds.) Informatik 2015, Lecture Notes in Informatics (LNI). Koellen Verlag (2015)
62. Mayring, P.: Qualitative Inhaltsanalyse, 11th edn, Beltz (2010)
63. Johnson, P., Ekstedt, M.: Enterprise Architecture—Models and Analyses for Information Systems Decision Making, Studentlitteratur (2007)
64. Buckl, S., Matthes, F., Schweda, C.M.: Classifying enterprise architecture analysis approaches. In: The 2nd IFIP WG5.8 Workshop on Enterprise Interoperability (IWEI'2009), pp. 66–79, Valencia, Spain (2009)
65. Hanschke, I.: Strategisches Management der IT-Landschaft: Ein praktischer Leitfaden für das Enterprise Architecture Management, 3rd edn. Hanser Verlag, München (2013)
66. Matthes, F.: EAM KPI Catalog v.1.0, Technical Report, Technical University Munich, Chair for Informatics 19, München, Germany (2011)
67. Hauder, M., Münch, D., Michel, F., Utz, A., Matthes, F.: Examining adaptive case management to support processes for enterprise architecture management. In: International Enterprise Distributed Object Conference Workshops and Demonstrations (EDOCW), pp. 23–32, Ulm, Germany (2014)
68. Object Management Group: Decision Model and Notation 1.0—Beta 1 (2014)
69. Weske, M.: Business Process Management: Concepts, Languages, Architectures. Springer, Berlin (2007)
70. Taylor, F.W.: The Principles of Scientific Management, vol. 202, New York (1911)
71. Bruno, G.: Requirements elicitation as a case of social process: an approach to its description. In: Business Process Management Workshops, pp. 243–254 (2010)
72. Bruno, G., Dengler, F., Jennings, B., Khalaf, R., Nurcan, S., Prilla, M., Sarini, M., Schmidt, R., Silva, R.: Key challenges for enabling agile BPM with social software. J. Softw. Maint. Evol. Res. Pract. **23**(4), 297–326, June 2011
73. Benkler, Y.: The Wealth of Networks: How Social Production Transforms Markets and Freedom. Yale University Press (2006)
74. Granovetter, M.: The Strength of Weak Ties. Am. J. Sociol. **78**(6), 1360–1380 (1973)
75. Tapscott, D., Williams, A.: Wikinomics: How Mass Collaboration Changes Everything (2006)
76. Vargo, S.L., Maglio, P.P., Akaka, M.A.: On value and value co-creation: a service systems and service logic perspective. Eur. Manag. J. **26**(3), 145–152 (2008)
77. Schmidt, R., Zimmermann, A., Möhring, M., Jugel, D., Bär, F., Schweda, C. M.: Social-software-based support for enterprise architecture management processes. In: Business Process Management Workshops, pp. 452–462. Springer, Berlin (2014)
78. Surowiecki, J.: The Wisdom of the Clouds. Anchor (2005)
79. Plataniotis, G., De Kinderen, S., Proper, H.A.: EA anamnesis: an approach for decision making analysis in enterprise architecture. Int. J. Inf. Syst. Model. Des. **4**(1), 75–95 (2014)
80. Uschold, M., King, M., Moralee, S., Zorgios, Y.: The enterprise ontology. Knowl. Eng. Rev. **13**(01), 31–89 (1998)
81. Kang, D., Lee, J., Choi, S., Kim, K.: An ontology-based enterprise architecture. Expert Syst. Appl. **37**(2), 1456–1464 (2010)

82. Wagner, G.: Ontologies and rules for enterprise modeling and simulation. In: 2011 15th IEEE International Enterprise Distributed Object Computing Conference Workshops (EDOCW), pp. 385–394, Aug 2011
83. Azevedo, C.L., Almeida, J.P.A., van Sinderen, M., Quartel, D., Guizzardi, G.: An ontology-based semantics for the motivation extension to archimate. In: 2011 15th IEEE International Enterprise Distributed Object Computing Conference (EDOC), pp. 25–34 (2011)
84. Sandkuhl, K., Smirnov, A., Shilov, N., Koç, H.: Ontology-driven enterprise modeling in practice: experiences from industrial cases. In: Advanced Information Systems Engineering Workshops, pp. 209–220. Springer International Publishing (2015)
85. Gruber, T.R.: A translation approach to portable ontology specifications. Knowl. Acquis. 5(2), 199–220 (1993)
86. Uschold, M., Gruninger, M.: Ontologies: principles, methods and applications. Knowl. Eng. Rev. 11(02), 93–136 (1996)
87. Bürger, T., & Simperl, E.: Measuring the benefits of ontologies. In: On the Move to Meaningful Internet Systems: OTM 2008 Workshops, pp. 584–594. Springer, Berlin (2008)
88. Antunes, G., Bakhshandeh, M., Mayer, R., Borbinha, J., Caetano, A.: Using ontologies for enterprise architecture analysis. In: 2013 17th IEEE International Enterprise Distributed Object Computing Conference Workshops (EDOCW), pp. 361–368 (2013)
89. Antunes, C., Caetano, A., Borbinha, J.: Enterprise architecture model analysis using description logics. In: 2014 IEEE 18th International Enterprise Distributed Object Computing Conference Workshops and Demonstrations (EDOCW), pp. 237–244 (2014)
90. Noy, N., McGuinness, D.L.: Ontology Development, vol. 101. Knowledge Systems Laboratory, Stanford University (2011)

Chapter 7
A Framework to Support Digital Transformation

Oliver F. Nandico

Abstract This chapter proposes a lightweight enterprise architecture framework which serves the demands of enterprise architects being confronted with a digital transformation scenario with an agile development approach. This framework therefore emphasizes the adaptability and the possibility for propagation of change throughout the defined architecture instead of addressing all possible concerns of all stakeholders. The framework itself is roughly based on TOGAF 9.1 and uses definitions of its content metamodel and follows ADM.

7.1 Changed Role of IT and the Enterprise Architecture in the Times of Digital Transformation

"IT no longer supports the business, IT is the business". This subtle play on words shows the role change information technology underwent in recent years. There is a growing part of enterprises and organizations, where information technology is in the core of the business, where IT not just supports but enables new or enhanced offerings. We call this evolution—or better, looking at the time frame—revolution "Digital Transformation" or "Digital Business Transformation".

7.1.1 Changed Role of the Architect

"Digital Transformation" sets a new challenge for the enterprise architect: She has now not just to align the IT with the demands from the business but to enable and even invent new business opportunities. So the architecture capability of an organization gets an active part in shaping the business. The architect no longer sits in

O.F. Nandico (✉)
Capgemini, Munich, Germany
e-mail: oliver.f.nandico@capgemini.com

© Springer International Publishing Switzerland 2016
E. El-Sheikh et al. (eds.), *Emerging Trends in the Evolution of Service-Oriented and Enterprise Architectures*, Intelligent Systems Reference Library 111,
DOI 10.1007/978-3-319-40564-3_7

her office awaiting demands or requirements from the business, but is part of the leadership team shaping the new, digital transformed enterprise. The architect does not only change her role in respect to the business, but to the IT as well. She needs not to just support, but to enable IT development for Digital Transformation.

7.1.2   Services as Atomic Building Blocks of the Architecture

Though there is an upheaval in enterprise architecture from this high perspective, the general approach and toolset of enterprise architecture has not changed so much. The old virtues of service oriented architecture still prevail, as using services, i.e., self-contained pieces of work with a focused business purpose, as the atomic elements for any architecture. Today, service-oriented architecture is the generally accepted standard for architecture work, which does not need any further discussion. If one wants to describe, what an information system does, she or he will start with describing its intended services. There is still an ongoing discussion on the definition of services, their granularity, technical aspects like SOAP versus REST and the need for additional elements like events and triggers etc. Nevertheless, services as basic elements for architecture have prevailed.

7.1.3   Time Is the Most Limited Resource

Time is the most limited resource in the digital transformation scenario. Therefore, the deliverables of the architect has to be limited to the amount really providing benefit. Basically this leads to a way how to define the deliverables required from the architect: Think of the issue to be resolved for a successful digital transformation, and—by this, derive the necessary deliverable.

7.1.4   Agile Approach Necessary

But new drivers require adaption for architecture as well: Volatile and fast-paced markets require flexibility and adaptability of the IT solution and an agile as well as lean approach for its development is the state-of-art to achieve this.

Some evangelists of "Agile" utter criticism of the conventional way of doing architecture or doing architecture at all: It would create many use- and meaningless concepts, lead to the notorious "Big Design Up-front", which no one helps. Enterprise architects would impose rules and guidelines on the projects which prevent simple, practical solution and prevent quick adaption to change. To be honest, many of these reproaches are quite justified, looking at some enterprise architecture practices.

There are several techniques and methods used under the broad idea of agile. So we refer here to the Agile Manifesto (http://www.agilemanifesto.org/) as the common ground for these. I want to emphasize two aspects from this manifesto:

- Working software over comprehensive documentation
- Responding to change over following a plan

So there is a need to develop architecture together with the working software. When following an agile approach, it makes no sense to develop thoroughly a complete architecture for the whole enterprise and have it implemented with the Digital transformation program. Instead the architect starts with a coarse overall draft plan, to be detailed when the need for decision arises or the answer to change is requested.

7.1.5 Need for a Lightweight Enterprise Architecture Framework

This requires a light weight enterprise architecture framework, simple to use and to maintain, with a minimal set of essential architecture deliverables. With TOGAF 9.1 I see an architecture framework as "a conceptual structure used to develop, implement, and sustain an architecture". The main purpose of such a framework are to visualize solution and solution alternatives for the organizational and IT structure with their consequences, to marshal all the development work within a digital transformation program and to provide guidance for architecture decisions.

As an answer to this challenge I propose such an enterprise architecture framework to support Digital Transformation developed "under fire", i.e., in practice, for digital transformation programs.

7.1.6 Overview of This Article

In this article we first look at the specifics of a Digital Transformation program and the requirements it poses for the architecture work. After this we motivate the use of TOGAF as a foundation for the proposed lightweight framework, and analyze the drivers for its customization.

After this we show how to resolve the issues of the conventional enterprise architecture approach. This is followed by a more detailed description of the viewpoints of the proposed lightweight enterprise architecture framework. Here we differentiate between the enterprise or program level on the one hand, where the architect works in a coarse and programmatic way and the project level, where she works making decisions and deriving solutions.

We close with a reflection of the challenges an architect has to face in a digital transformation program.

7.2 Digital Transformation and the Consequences for a Respective Framework

To address the right issues for the proposed framework, we have to define the term "Digital Transformation" one step further. So, when we speak of "digital transformation" of an enterprise, we mean the change of this enterprise with the intention to provide a new or significantly enhanced offering to its customers, where this new offering or enhancement is based on information technology as key enabler or even part of the offering.

The new or enhanced offering will comprise new or enhanced services, products or both. Basically, the enterprise uses "digital transformation" to take the opportunities the development in information technology provides. Initiatives for digital transformation are centered on envisioning new business models, customer experience and operating models with their respective operational processes.

The following examples may clarify the Digital Transformation scenario:

- A passenger transport carrier changes to a mobility service provider where it now not only provides own transport services but in addition brokers other provider's services to get a passenger from point A to point B.
- An apparel manufacturer changes from ready-made clothing to custom-made where its clients use augmented reality to make their orders.

Though the digital transformation scenario needs to and will change the IT of the respective enterprise drastically, it is totally business driven, i.e., the drivers are rooted in the business, and business decides on executing the "digital transformation" scenario. As a result of this transformation, the mission, vision and operating model of the enterprise may change radically, and so will the business architecture of the respective enterprise. So the architecture work for digital transformation is strongly governed by the business context, much more as it has been in the past.

This differs totally from the "IT only"-transformation programs we used to see. What has been and still is called a SOA program, i.e., a transformation program to implement service oriented architecture, was a change more or less affecting the IT only. Even when IT consultants talked about to provide better alignment with the IT and more IT flexibility, those programs aimed essentially at IT rationalization and modernization. As a result, business appreciated the pursuit of IT, but did not realize IT as anything like an enabler for new business opportunities.

This has now changed. For many industries IT has or will become a vital if not the only part of the business. With the introduction of smart phones, wearables and the internet of things these digital transformed businesses find their counterpart at the consumers' side.

Of course, there is still the need to move and transform physical objects. But with the progress which 3D printing has made, one can imagine that even this may shift to information technology.

So what are the implications of digital transformation for service-oriented and enterprise architecture?

Clear business orientation: The architect needs to design for business outcome and business value. She has to look explicitly for new business opportunities and by this exceeds the traditional limitations of enterprise architecture. So a framework for digital transformation has to reflect this clear business orientation.

Building something new: Digital transformation is about building something new, business and IT-wise. Therefore new systems are preferred to the adaption of legacy systems: Even if the enterprise wants to keep legacy systems, they must not slow it down or create the need for half-baked compromises. A slash-and-burn approach will usually be the approach of choice. It may make sense to keep legacy back-office systems in strongly regulated business areas, e.g., finance, if they go along with the new business models and operational processes. But the architect has to be careful: Do not perpetuate big legacy systems at the end of their life cycle instead of replacing these by truly standardized systems or outsourcing the whole process.

Integration beyond organizational boundaries: In the era of digital transformation enterprises are finally no longer islands in the vast ocean of the internet, with a defined boundary between land and sea. These boundaries between enterprises get fuzzy, if not totally dissolved. Enterprises, organizations and even consumers communicate, share information and interact by means of information technology services. These IT services may be used inside of an organization in the same way as provided to the outside. The value-adding process in the digital sphere is based on using some IT services of some providers, combine and augment them and provide them to respective consumers, beyond any organizational or other boundaries. So a framework for digital transformation is truly service oriented.

When we speak of "Digital Transformation", we assume a **digital transformation program** for the whole or at least a bigger part of the enterprise, i.e., the organization in scope. This program usually is broken down into manageable parts, i.e., **projects** within this program. Therefore I differentiate in this article between the work of the architect at program or enterprise level and at project level.

7.3 The Lightweight Enterprise Architecture Framework—A Very Focused Customization of TOGAF

There are many enterprise architecture frameworks the architect can chose of: From the ancestor of all of them, the Zachmann framework to the widely accepted standard TOGAF, the architecture framework of the open group and some more which come with various tools the architect may use.

All of these architecture frameworks are quite comprehensive and it will cost much effort and time to follow them thoroughly. Obviously, there is no use in defining just another framework. As TOGAF is the general accepted standard for enterprise architecture, we follow in this description the TOGAF definitions. This proposed framework is not totally new, but orients towards and makes use of TOGAF, especially of its Architecture Development Method (ADM) and its content metamodel. So the framework proposed here is basically a very focused customization of TOGAF.

Taking TOGAF literally and to the full extent of its described deliverables, there is always the risk of interpreting it in a way we want to avoid. In the digital transformation scenario, it is of no use to define the whole target architecture in terms of business, information system and technology architecture after obtaining a statement of architecture work, which is based on the architecture vision. (TOGAF 9.1). As learned from some digital transformation projects, the enterprise architect needs her assignment based on the business strategy, the underlying business ideas and the business objective. The enterprise architect may already contribute to this according to the opportunities IT may provide to the business ideas. An architecture vision and a coarse architecture action plan—an extension to TOGAF for Phase A proposed here—aligned with stakeholders is all what is needed to start a digital transformation program. The actual architecture change takes place in the projects of this digital transformation program. The digital transformation program can and will not wait for a comprehensive architecture designed by the omniscient and omnipotent enterprise architect.

The TOGAF ADM proposes a preliminary phase, which serves for customization, tool selection etc., basically the preparation of the "TOGAF architecture project". Though I think some preparation is always useful, the proposed lightweight enterprise architecture framework does no reference this phase explicitly. The considerations of this article together with implementation for an actual digital transformation program more or less follows the idea of this preliminary phase.

7.4 Drivers of Digital Transformation Provide a Foundation for Architecture Guidelines and Principles

Both, the architecture vision and the architecture action plan, have to reflect the drivers of a digital transformations program. Instead of trying to resolve every architecture issue up front, the architect shows and prioritizes the forces which will rule her decisions. These drivers set constraints for the architecture approach und thus form architecture principles and guidelines. If the architect does not consider these drivers, her work may be useless or ineffective at best. On the other hand, just stating these some architecture principles without basing any decisions on it, turns those principles into useless documents.

The following main drivers of digital transformation program will lead to a good foundation of principles for governing the architecture work:

- Reducing time-to-market
  There may be two archetypical situations within the digital transformation scenario. Either the enterprise is ahead of the completion, the first with the new business idea and wants to make most of this competitive advantage or the enterprise has to catch up with competitors which already make their business in the digital sphere and threaten the enterprise's market position. Either way, there is a very strong pressure to reduce the time for implementation of the business idea and to provide the new services to the customers, i.e., the time-to-market. But there is a caveat: One may fail, if the new offering is technically immature or provides security risks. So the architect is caught in the middle: Architecture design must not slow down the implementation of the business idea, but has to show a robust, yet flexible structure of the solution, which the core business deserves.
- Flexibility, robustness and responsiveness
  Digital transformation requires services at any time, at any location, with any device with minimal response time. Usually, this is nothing one can build as an addition to the existing process and IT landscape, but has to be designed as an intrinsic element. Separation of model from view, flexible routing concepts and online solutions instead of batch processing are old design virtues, in memory solutions and mobile app framework new technology solutions to be considered.
- Security and reliability
  It is hard to gain the user's trust and it is easily lost. Therefore, all architecture decisions have to put in security and the users' requirements on data privacy in the first place. But security comprises the aspects of availability at all times as well. When information systems are not available, users will look for another—and may be lost. So investment in robustness and resilience of IT solutions are investments in customer loyalty.
- Small, single purpose services as elementary building blocks
  Functional flexibility calls for single purpose, mutually independent, highly decoupled and context free services, a concept termed "micro services". For creating options within the architecture, respective IT services shall be able to be deployed independently and easily switched on or off during run time. This obviously requires new ways of automated deployment and operations. Of course, this concept of small, single purpose services may increase cost as well, and the architect has to balance this with the business value for meeting the "window of opportunity".
- Focus on standardization and business value of information systems and their services to be added:
  To build up the solution as fast as possible one has to rely on already available services, especially for infrastructure, platform and integration. Those information systems and platform services may be provided by external service provides, as a cloud services, by package based applications or even legacy

applications. They have to comply with common standards, so they can be orchestrated easily within the solution.

- Complete real-time automated end-to-end processes
  Within the digital transformation scenario we build new end-to-end fully automated processes, which work real-time. There are no over-night batch processes but direct and immediate reaction of the whole application landscape and update of the complete information status.

The architect considers these drivers for the architecture decisions she makes. This consideration leads to principles and guidelines for the architecture to be defined.

7.5   Issues to Be Addressed for Digital Transformation by a Lightweight Enterprise Architecture Framework

A lightweight framework for digital transformation has to address the following issues:

Issue: What has the digital transformation of the enterprise to achieve?

This issue is about, why we actually want to transform the business. There is a business idea and some business goals with an underlying business strategy, and set business objectives according to this business strategy. When looking at business transformation, opportunities the IT provides trigger the business idea, at least to some extent. So there is already a close mutual interference between business and IT at the starting point of digital transformation.

There may be some additional business constraints: the given time frame, geographical scope, a general functional scope, the target operation model etc. If the legacy IT landscape is of any importance, a coarse heat map or gap analysis is also part of these considerations.

All together, this forms the **Architecture Context** for digital transformation. The architect is well advised, to clarify this context with all the stakeholders before a first step of his work. This clarification comprises also her formal assignment, the statement of architecture work as referred to in TOGAF: "verifying and understanding the documented business strategy and goals" (p. 71). The context may be mostly vague in some parts in the beginning. Some aspects of this context may even change during this digital transformation endeavor. So the documents for describing the architecture context shall allow tracing changes of it and the decisions which triggered the changes.

As the digital transformation scenario is focused on implementing something new, the architect does not take the legacy business strategy much into consideration. If this is part of the acknowledged Architecture Context it is usually referred to as an add-on to the target business strategy and thus part of it.

The benefit in compiling this architecture context is to set a goal for the digital transformation in order to align all stakeholders. As digital transformation is usually a risky and costly engagement, it is imperative to rally everyone around the set goals.

From a TOGAF customization perspective we formally add Architecture Context as an architectural input. We do not assume that it will be fully clarified or in any way constant for the digital transformation program, not even for an iteration. Quite the contrary, it will and has to change according to market experience and demand, customer feedback, competitive situation etc.

Issue: What is going to be realized?

Derived from the Architecture Context, the architect creates an architecture sketch. She looks for overall value chain of the enterprise, the business and IT capabilities required to realize this value chain and orders all of this in some reasonable way, in so-called domains. This may already include some preliminary decisions on in- and outsourcing, and thus on the scope. This will show a way to make the achievement of goals for the digital transformation program feasible. All in all, this results in an **Architecture Vision**.

The benefit of an architecture vision is the alignment of the architect and the stakeholders of the transformation on the future structure of the transformed enterprise. By this the architect's assignment from the architecture context gets clarified, but there is no need for a formal statement of architecture work—which is in practice rarely issued.

Looking at the TOGAF customization, we see the architecture vision as the first deliverable of the enterprise architect, not as a means to get the statement of architecture work signed off. The statement of architecture work is therefore skipped. Following the principles of the agile manifesto, we see no value in a formal assignment for the architect. The respective stakeholders should trust her in realizing the goals and objectives as compiled in the architecture context. This follows the statement from the Agile Manifesto "Customer collaboration over contract negotiation", where the digital transformation program may be seen as the architect's customer.

Issue: What are we going to do to realize the architecture vision?

After clarification on the architecture vision, there is a need to describe on a high level the necessary measures to get this architecture vision into reality. This comprises two deliverables: From the overall drivers of digital transformation and the architecture context, we derive business and architecture principles as guidelines for the actions to be taken and for future architect decisions. So, they form the guide rails for the digital transformation program. These architecture principles applied to the domains in the architecture vision gives a foundation for measures to be taken realize the architecture vision. The description of these measures has to be rather coarse to leave room for options and reaction to change. They are to be detailed in individual projects to be started within the digital transformation program. So the architecture principles together with the defined measures for realizing the architecture vision form the **Architecture Action Plan**.

The architecture action plan is the architect's input to the digital transformation portfolio and roadmap. Of course, the management of a digital transformation program has to take some other aspects beyond the architectural aspects into consideration when deciding on project portfolio and roadmap, like the availability of knowledgeable resources, investment plan etc.

These three activities, compiling the architecture context, deriving the architecture vision and the architecture action is all what the enterprise architect needs at the beginning of a digital transformation program. Anymore details shall be left to the individual projects/initiatives or later architecture decisions to be made at the right time. So architecture vision and architecture action plan have to be coarse and flexible. Business may change at short notice, and we want—true to the agile manifesto—embrace change.

Especially there is no use in elaborated design before the actual implementation starts. It will just slow down the program, and any decisions taken will be too early and will not pass the test of time.

Of course, the architect bases architecture vision and architecture action plan for a digital transformation always on business considerations and business architecture. So there is no sense putting business in opposite to IT, we need a holistic view. And we have to keep in mind our starting statement: IT is the business.

TOGAF 9.1 lacks something like an architecture action plan, and proposes an architecture roadmap instead. But this architecture roadmap needs a defined architecture as prerequisite. This is too detailed for the approach we want to follow here.

Issue: What is the contribution of a certain project to the realization of the architecture action plan?

When the digital transformation program starts, i.e., the defined project portfolio gets worked off; there is a need to provide guide rails for each project. According to the defined scope of each project, the measures from the architecture action plan defined for this respective scope become part of the project's backlog.

This is formalized by an **Architecture Contract** or an **Architecture Outline** for a certain project. This deliverable documents the contribution of the project to the architecture action plan.

By putting together the architecture outlines of the projects with their degree of implementation the enterprise architecture is able to value the degree of realization of the architecture vision.

TOGAF 9.1 thinks of an overarching architecture project, which then starts certain work packages to implement the proposed architecture. In practice, the enterprise architect almost never starts projects. Usually, she is an advisor to the portfolio management when defining individual projects. But to ensure the alignment of the projects within a digital transformation program, the architecture outline is needed. Otherwise all the sprints within the projects will only result in a "caucus race" (Lewis Carrol, Alice in Wonderland). So the Architecture Outline in the sense used here is an add-on with respect to TOGAF 9.1.

Issue: What functionality does a project implement?

We assume that the individual projects within a digital transformation program will follow an agile approach. Given that there will be the need for adapting the digital transformation program at short notice. There is no alternative to it. So user stories provide functional requirement which populate a projects' backlog. These functional requirements have to be fulfilled by atomic building blocks, the services. When we look at these from a business perspective, we see them as business services.

A business service is described on the one hand by a triple: its business goal, i.e., the purpose of a service, the business role as the role of the actor of this service and the activity this actor executes. Another viewpoint for a business service is the business objects as consumed input of a business service and created or transformed business objects.

Linking the business services by their provided respectively consumed business objects shows the dependency between the services. This forms a special viewpoint for business services, the business collaboration view.

From the required business services the project architect will derive the required information system services and the platform services needed to support these. She will map technical or "non-functional" requirements to these derived information system services. As this takes place under the imperative of *digital* transformation we can assume a one-to-one relationship between business and information system services.

As a result the project architect maintains a **Conceptual View** of the services to be implemented in the course of the project. This conceptual view consists of the cross reference of business services, information system services and platform services and the business collaboration view.

With this conceptual view, the project keeps track on already implemented services and its backlog of functionality still to be implemented.

This conceptual view customizes TOGAF 9.1 strongly. This approach does not go for the whole business architecture, as we only need the required business services. How these business services collaborate the business collaboration view shows. A digital transformation scenario focuses on the application architecture. Therefore the information system services form the basis for the further architecture considerations.

Issue: How do the services work together?

The heart of the architecture work in a project is the grouping of the information system services within the scope of a project. From the business collaboration view the project architect sees the dependencies between certain business services, which result in dependencies of information system services. Architecture principles, the domain structure as defined in the architecture vision and the joint utilization of business objects result in an ideal structure of components. The architecture outline of the respective project sets scope and guidelines for this blueprint.

By the terms of digital transformation we think of the application structure reflects directly the business structure. The project architect derives application components based on the business structure. So we get a logical view on application components.

A project architect orders the platform services required to support the application services as well and thereby creates the logical view on technology components.

Both, the logical application component and the logical technology component view, form the **Logical Component View**. Its purpose is to show the ideal intended architecture and the effect of architecture decisions to it.

Again, this is a very strong customization of TOGAF 9.1, focused on application architecture within the scope project and taking the necessary technology architecture in consideration. This logical component view is seen as a minimal viewpoint, a project architect may need to derive further viewpoint when special concerns of stakeholders become apparent.

Issue: With what does the project realize the logical components?

Considerations from the architecture context, like procurement of services, like using cloud service, utilizing a software package or even continue legacy applications lead us to the actual components to build. For the majority of required services and logical components these will be provided by either commercial-off-the-shelf (COTS) software packages or services provided from the cloud. As time-to-market is the strongest driver within the digital transformation scenario, newly build services and their respective components have to be limited to the really necessary ones. Ideally, we follow a more or less mesh-up approach for the implementation.

So the architecture has to decide between different solution alternatives using different combinations of software packages and cloud solutions—and possible different ways of integrating them. This leads to a number of architecture tradeoff decisions based on the ideal structure given by the logical component view.

The resulting **Physical Component View** is the IT blueprint and will serve to resolve deployment and operations issues. It provides the project with a concise and consistent view on the decisions which are meant to last. Of course, this view will show areas where adaption to change is still possible, due to built-in options.

So this physical component view provides the project a concept for a stable base on which the project can build upon.

Again, for the physical component view we focus on application and technology architecture in combined way, which is different from the approach TOGAF 9.1 takes.

So within a project, the project architect creates and maintains three basic views for the project: the conceptual, the logical and the physical view, with the architecture outline as the starting point.

Issue: How does the enterprise architect execute architecture governance?

The enterprise architect is responsible for maintaining and adapting the architecture vision and the architecture action plan, tracking the progress the projects make towards realizing the architecture vision and the carrying through of the architecture action plan. These are the competencies by which the enterprise architect executes architecture governance.

To get information on the progress the projects within the digital transformation program have made, the enterprise architect organizes **Architecture Reviews**.

This review checks how far the project has implemented the architecture outline but also derivations from this architecture outline and need for adaptation of the architecture action plan and/or the architecture vision.

Business reasons, outside triggers or any other reason may pose the need for change on the architecture vision and—consequently—on the architecture action plan. This, of course, triggers change within in the architecture outline of the projects and change in the projects, especially in the project's architecture views and implementation. So another part of the architecture governance is the tracking of applied change throughout the digital transformation program.

The benefits of architecture governance are quite clear. Only with effective architecture governance the architecture vision will be realized. If there is no architecture governance, it will be difficult to apply change of the digital transformation program to all the projects.

7.6   The Viewpoints of the Lightweight Enterprise Architecture Framework

According to the issues the enterprise architect has to address, we defined viewpoints and deliverables of a Lightweight Enterprise Architecture Framework for Digital Transformation. The general maxim for this is "as minimalistic as possible". For this framework we only define those deliverables which we see as mandatory. Instead of putting effort into more and more viewpoints, we want to keep the existing set current and consistent and propagate any change directly. Effort is best invested in keeping the core viewpoints up to date, instead of creating a big set of viewpoints one cannot maintain all of them.

There may be a need to generate further viewpoints addressing certain concerns of specific stakeholders. These additional viewpoints shall be based on what is defined by this minimal set. If so, we advise to use them only as snapshots with temporary character of the currently defined architecture.

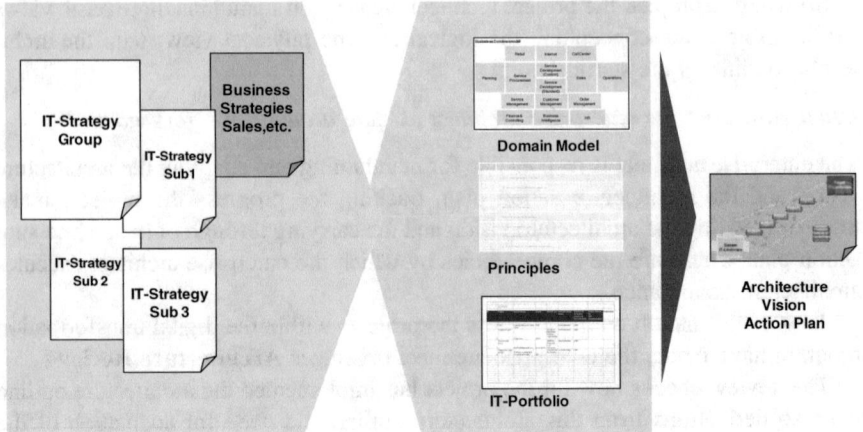

Fig. 7.1 Deriving architecture vision and architecture action plan

7.6.1 Viewpoints at Enterprise Level

There is a challenge for the enterprise architect: How to align the need for an overall architecture definition at enterprise level with the implementation design the individual projects do and where the actual change occurs? The proposed way for this is, to keep the architecture vision at a coarse level to give flexibility to the projects but to provide rail guides to orchestrate the pursuits of the projects towards the common target architecture (Fig. 7.1).

7.6.1.1 Defining the Goal: Architecture Vision

Goals and deliverables

Objectives of Phase "Architecture Vision" according to TOGAF are to "develop a high-level aspirational vision of the capabilities and business value to be delivered as a result of the proposed enterprise architecture" and to "obtain approval for a Statement of Architecture Work that defines a program of works to develop and deploy the architecture outlined in the Architecture Vision". Unlike TOGAF, we see the architecture vision already as assigned architecture work and as the answer to the business request for digital transformation.

The purpose of the architecture vision is the creation of an overview and to set guidelines for all projects within in a digital transformation program in order to achieve its business goals.

A domain model provides a business driven framework to the envisaged change of the IT landscape. So the overall change is detailed by ordering the description of the future state and the required business capabilities to the domains of the domain model.

All the information from the architecture context and the architecture principles as derived from this architecture context, provide the guidelines for the proposed architecture visions.

Input

- Architecture Context
  Business idea, business drivers, business mission & vision and business objectives give us the business rationale for the digital transformation endeavor. The target operating model and the required business capabilities enables the design of the architecture visions.

Activities

- Derive and align architecture principles
  An architecture principle is an approach, a statement or belief that drives the architecture development. In this step, we take the architecture context and derive those statements from them which are significant for the target architecture.
- Compile and align the domain model
  We define domains based on the top level value chain and top level required capabilities to structure the information system landscape.
  "A domain is a business area of focus, interest, study, and/or specialization. It is an area for which a body of knowledge exists. Sometimes this is just a tacit body of knowledge but it is nonetheless a body of knowledge. All other things aside, domains are the primary tops of a system" (Coplien, Lean Architecture). So we see a domain as an abstract that describes an area of knowledge, rules, policies, views etc. and provides an ordering framework for architecture artifacts. As we assume that business and information system aspects are consistent in a digital transformation scenario we define only one domain model for both aspects.
  We align the compiled domain model with the respective stakeholders in order to specify domains in a way so that we can map the required business capabilities to a domain. This mapping characterizes the domains further (Fig. 7.2).
- Create and align the architecture vision for the digital transformation program
  According to high-level demands and the architecture principles for each domain the target architecture of each domain is described as some top-level logical components with an outlook of the planned target architecture in each domain.

Output

- Domain Model
  A graphic representation of all the domains with a concise and consistent description of the domains.
- Architecture Vision
  One or more general statements per domain and a view of proposed top-level logical components on the pursued target architecture as a result of high-level requirements, business objectives and goals.

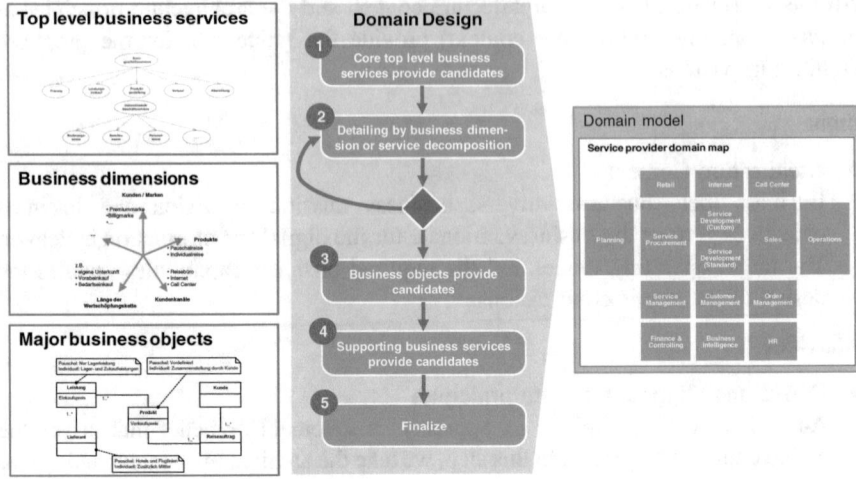

Fig. 7.2 Creating domain model

7.6.1.2 The Enterprise Architects' Instrument of Effectiveness: Architecture Action Plan

Goals and deliverables
The architecture action plan shows, how the target architecture gets implemented, i.e., what the measures to introduce new capabilities are and how to address the stakeholder concerns.

Basically the architecture action plan outlines a program of works to develop and deploy the architecture by the digital transformation program. For upcoming engagements and programs, the architecture action plan provides guidelines how to contribute to the overall architecture roadmap to realize the architecture vision.

Input

- Architecture Context
- Architecture Vision.

Activities

- Create and align the architecture action plan for digital transformation
  According to the envisioned target architecture for each domain the measures to be taken for implementation are described in general terms.

Output

- Architecture Action Plan.

7.6.2   Viewpoints at Project Level

The actual change, the implementation of the digital transformation, happens in individual defined projects within the transformation program. So the projects' architect does the detailed architecture work which results in an implementation blueprint. Obviously, there is a need to connect the architecture in each project to the architecture vision and the architecture action plan. These connections are the architecture outline, which defines the contribution of each project to the overall architecture vision, and architecture reviews, which checks if the architecture created in the project and the architecture at enterprise level are still in sync. These architecture reviews may call for change of either the enterprise architecture or the project architecture or both.

Change may occur on enterprise level as well and the architecture vision and subsequently, the architecture action plan have to adapt. This may lead to some change for the architecture outlines for some projects. To propagate change quickly, effective and thoroughly, the enterprise architect has to carefully maintain the relationships and traceability between architecture vision, architecture action plan and architecture outlines. This requires automation and tool support.

7.6.2.1   Giving Rail Guides for the Projects: Architecture Outline

Goals and deliverables
The architecture visions shows a coarse picture of the target architecture, the architecture action plan lists the measures to be take to achieve it. The individual projects of the digital transformation program shall contribute to the architecture action plan and thereby to the realization of the architecture vision. To ensure the projects moving in the right direction, i.e., as given by the architecture vision and the architecture action plan the enterprise architect aligns with the project at the very first stage.

By the given scope of a project, the architect "places" the project within the domain model, so that architecture responsibility is clarified and which statements from the architecture action plan apply. By the comparison of the proposed projects scope and goals with the architecture action plan, the architect recognizes, how the project may contribute in principle to the architecture action plan and to what extent. The architecture outline corresponds to the architecture vision in the sense of TOGAF (Fig. 7.3).

Input

- Architecture action plan
  The current architecture action plan as enterprise architecture has defined.
- Project proposal
  We assume that a project proposal comprises at least the business objectives, top level requirements, the scope and a coarse solution idea for this project.

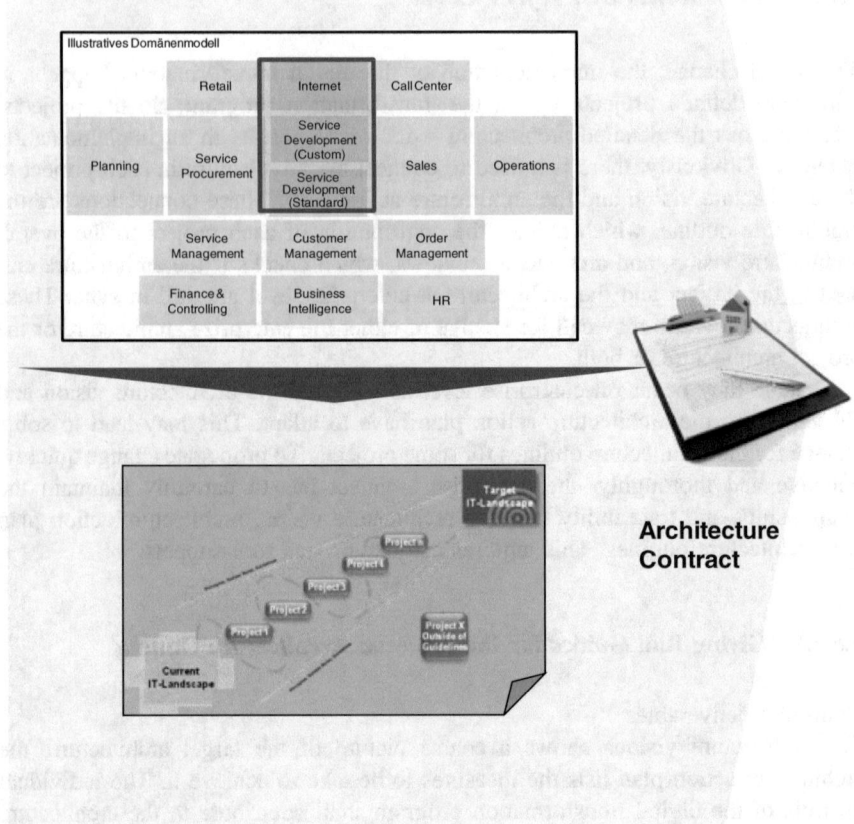

Fig. 7.3 Define architecture outline

Activities

- "Place" the engagement in the domain model
  Based on the given scope and solution idea of a project, the architect maps this
  engagement to the domains affected by it, i.e., she "places" a project in the
  domain model.
- Determine architecture requirements and recommendations for the project
  Given by the placement within the domain model, the architect determines the
  relevant statements from the architecture action plan which applies to this
  project. Thereby, she describes, how the project shall contribute to the implementation of the architecture action plan.
- Align and sign off of an architecture outline
  We align and discuss the architecture requirements with the project stakeholders, relate how they can be fulfilled by the project, with its business goals given
  and how the enterprise architect will support.

The agreed architecture requirements and recommendations, i.e., the architecture outline of the project form the architecture contract, which will be formally signed off.

Output

- Functional Placement
  Mapping of the scope of the project to the domain model
- Architecture Outline
  The architecture outline is an agreement of the project with enterprise architecture. It consists of architecture requirements and recommendations as given by the scope of the project and the applicable parts of the architecture action plan.

At the level of a single project, architecture becomes concrete and lead to an implementation blueprint. But even when the sequence of phases given here may imply an order, it is not meant this way. Basically the project's architect starts with a rather "blurred" model. This is based on a few services, a very coarse logical view and a quite incomplete physical view. With the discovery of more and more services, the architecture becomes clearer. From decisions at the logical and physical level, implications are fed back. As the physical architecture is meant to be implemented, there is feedback from implementation to the architecture as well. So the proposed deliverables are not in any way solid blocks resistant to change, but flexible and adaptable networks of decisions. The challenge for the project's architect lies in maintaining the relationships within this network and the right propagation and channeling of change.

So the viewpoints and phases proposed here are a combined approach what TOGAF ADM sees as the phases B, C and D.

7.6.2.2 What the Project Implements: Conceptual View

Goals and deliverables
Starting with the architecture context, the project's architect wants to derive detailed functional and non functional requirements to determine the detailed target architecture within the scope of the project. From the required new business capabilities, the other demands given by architecture context and by interaction with the stakeholders, user stories will be derived. Based on them, the required business services are determined. We see the business services as atomic in the context of the projects, i.e., they refer to one activity of one role for an elementary business goal. For these business functions we determine the business objects which are in- and/or output of the business functions.

The business collaboration view shows the dependencies between business services based on a consumer-producer relationship of business objects between business services.

Within a digital transformation program business services shall translate directly into information system services as we assume that any functional requirement will be fulfilled by some function within IT. The relationship between business services based on passing business objects translates into a flow of data entities between the respective information system services.

For the derived information system services the architect looks for required platform services to support them.

Input

- Architecture context
- Architecture vision
- Architecture outline.

Activities

- Detail requirements
  By appropriate means and tools, the top level requirements are detailed.
- Determine business services and business objects, detail them to the right granularity
  From the functional requirements on the appropriate level, the architect derives required business services. Business services are atomic elements of business behavior, characterized by one goal or purpose, one activity and one role executing this activity. Input and output of a business service are business objects (Fig. 7.4).
- Marshal business services
  We map the business services to the appropriate domains. We derive how the business services work together: Business services are related to each other by the business objects one business service produces and other business services consume. The viewpoint for this is the business function collaboration view (Fig. 7.5).

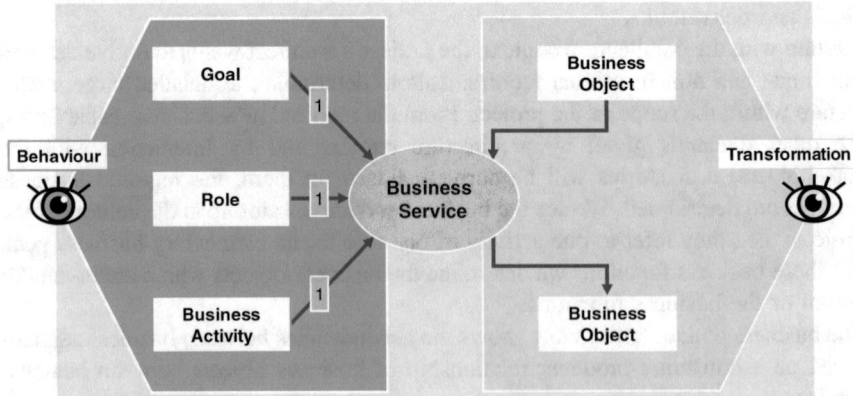

Fig. 7.4 Business service

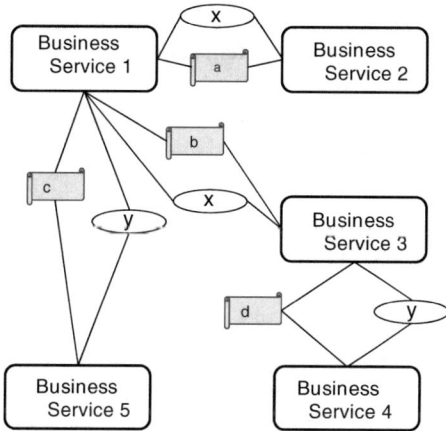

Fig. 7.5 Business collaboration view

- Derive required information system services
  We derive information system services from the business services in scope: Information services will implement the activities of business service in IT as we assume full automation of business services as the target state of digital transformation (Fig. 7.6).
- Derive required platform services.

Output

- Requirements and user stories
  Functional and non-functional requirements at the right level of detail
- Business services

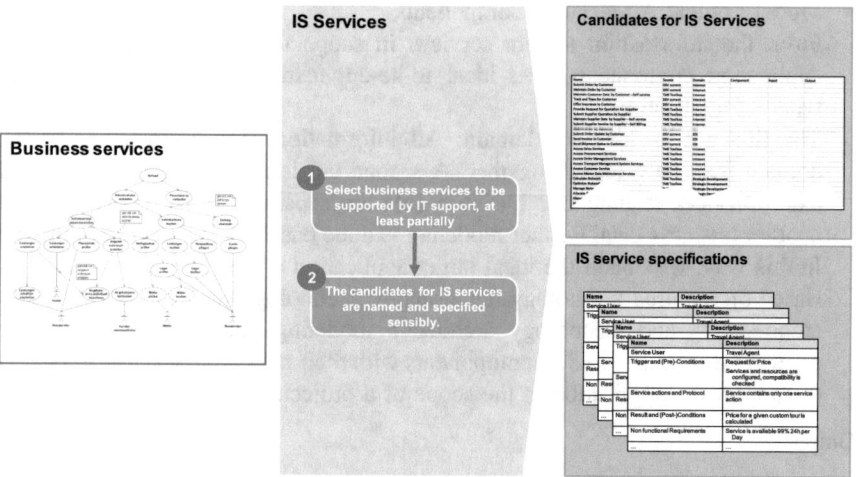

Fig. 7.6 Deriving information system services

- Business service collaboration view
- Information system services
- Platform services
- Cross reference business service—information system services—platform services.

7.6.2.3   How Services Work Together: Logical View

Goals and deliverables
The architect models and composes an ideal structure of application and technology according to her architecture thinking, the architecture principles and further modeling criteria like grouping services of only one domain to application domain or services working on the same or related data.

Input

- Architecture context
- Architecture vision
- Architecture outline
- Business services
- Business service collaboration view
- Information system services
- Platform services
- Cross reference business service—information system services—platform services.

Activities

- Model Logical Application Components
  From the information system services in scope the architect models logical application components, i.e., ideal to-be-applications/components or "proto-types of application".
  The first cut is done by domain: All information services in a domain/sub domain may be grouped together. A second cut is done by usage of business data: Services reading one business data are separated from services reading another business data. Further modeling is done based on architectural thinking. In this way, we obtain the ideal target application architecture (Fig. 7.7).
  Based on required technology functionality, technology standards and reference architectures and technology architect's modeling, the project's technology architect determines logical components with their respective services to be used to support the application in the scope of a project.

Output

- Logical Components.

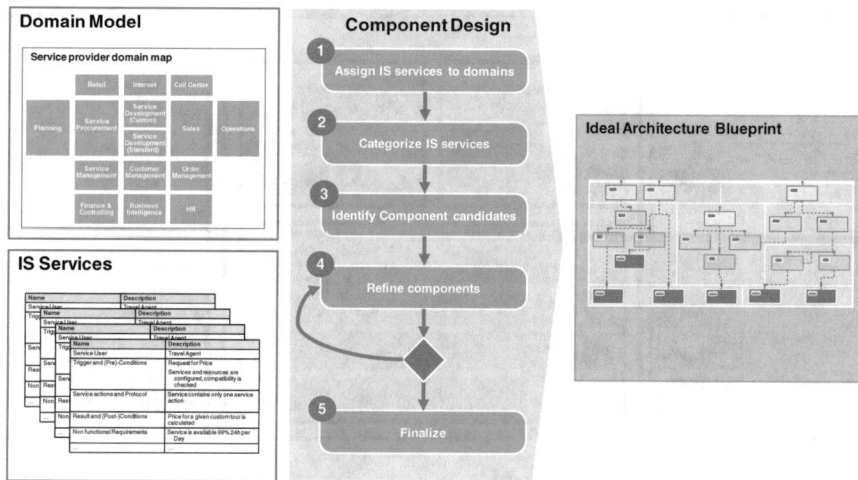

Fig. 7.7 Define logical components

7.6.2.4 With What the Services Get Realized: Physical View

Goals and deliverables
The architect takes the logical view of his project and derives different solution alternative from this by evaluating the use of COTS software and cloud services covering the logical components and their respective services. So there is a fit gap analysis for the different solution alternatives.

Input

- Logical Components.

Activities

- Derive solution alternatives
  We compile prospective solution alternatives, by considering buy or make alternatives or even the right use of legacy systems. A market analysis and a exploration of the intended future IT ecosystem of the digital transformed enterprise leads to available services, as cloud services and applicable COTS software. As time-to-market is one key success factor, digital transformation should make use of as much as possible of already available services. A solutions alternative is then a combination of software products which covers the logical component structure at least partially and complies with the architecture outline of the project.
- Conduct fit gap analysis and decide on solution
  For each derived solution, the project architect checks the coverage of the logical architecture in terms of the information system services provided, and

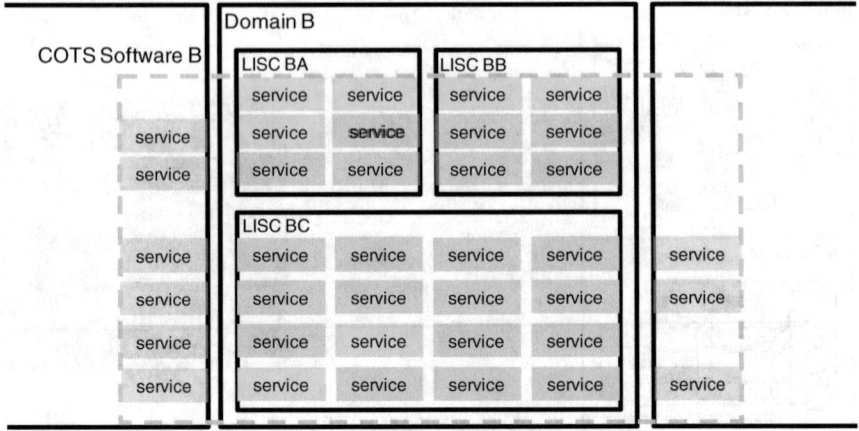

Fig. 7.8 Fit gap analysis at service level

detailed requirements fulfilled. The resulting gaps are assessed and decisions have to be made how to close or mitigate them—or accept them (Fig. 7.8).

Output

- Physical Components.

7.6.2.5 How to Spread Change: Architecture Governance

Goals and deliverables
The architect checks the actual project architecture against the architecture outline and compiles the deviations. From these deviations changes or amendments of the project architecture, the architecture outline or even the architecture vision and action plan may result which has to be fed back to the respective deliverables and propagated accordingly.

Input

- Architecture outline
- Project architecture: Conceptual, logical and physical view.

Activities

- Review project architecture
  Together, the enterprise architect and the project architect assess the current conceptual, logical and physical view of the project's architecture and check if the constraints and assignments of the architecture outline are met. If not, they state a deviation for any gap they recognize. In addition, they look for decisions the project architect has made in detailing the project architecture and decide if

and how these decisions influence the architecture outline. This will be formulated as amendments.
- Execute architecture change
  For any deviation a decisions has to be made how to mitigate it. This defined mitigation may cause change for the project architecture or amends the architecture outline. Amendments to the architecture outline may result in changes for the architecture vision and architecture action plan. The enterprise architect executes this change and propagates it accordingly.

Output

- Project architecture deviations
- Architecture change.

7.7 The Challenge for the Architect of a Digital Transformation Program

As stated above, the proposed framework reflects the key issue of digital transformation programs: Provide the flexibility and adaptability to actually enable, even embrace change on architecture. On the other hand, it is stated "architecture is about decisions that last". But if we think of architecture as solid blocks, the waves of change will destroy them over time. Therefore, architecture has to be done in a way that it can absorb change, propagate it to the right point and by this, control it.

There are risks, though. Changes may come in an uncontrolled way, as a tidal wave and overwhelm the architect as well as the architecture. Then measures have to be taken to canalize the flood, which usually results in drawing a stable baseline for the architecture. Concurrent and contradicting changes may cause to the architecture to "swing". By aligning the architecture changes more closely to the business objectives and the goals of the digital transformation, the architect can halt this and enable control again.

References

1. The Open Group: TOGAF™ Version 9.1. The Open Group (2011)
2. Schwaber, K., Beedle, M.: Agile Software Development with Scrum. Pearson Education/Prentice Hall (2002)
3. Hoogendoorn, S.: Das kleine Agile Buch. Pearson Deutschland (2013)
4. van't Wout, J., Waage, M., Hartman, H., Stahlecker, M., Hofman, A.: The Integrated Architecture Framework Explained: Why, What, How. Springer-Verlag (2010)
5. Engels, G., Hess, A., Humm, B., Juwig, O., Lohmann, M., Richter, J.-P., Voß, M., Willkomm, J.: Quasar Enterprise, Dpunkt-Verlag (2008)

6. Bente, S., Dr., Bombosch, U., Langade, S.: Collaborative Enterprise Architecture: Enriching EA with Lean, Agile, and Enterprise 2.0 Practices. Morgan Kaufmann Waltham, Elsevier (2012)
7. Op't Land, M., Proper, E., Waage, M., Cloo, J., Steghuis, C.: Enterprise Architecture: Creating Value by Informed Governance. The Enterprise Engineering Series. Springer-Verlag, Berlin, Heidelberg (2009)
8. Coplien, J.: Lean Architecture: For Agile Software Development. John Wiley and Sons Ltd., Chichester (2010)
9. Westerman, G., Bonnet, D., McAfee, A.: Leading Digital: Turning Technology into Business Transformation. Harvard Business Review Press (2014)

Chapter 8
A Two-Speed Architecture for the Digital Enterprise

Oliver Bossert

Abstract In all customer facing businesses, time to market is a key differentiator. The quicker a business is, the more successful it is likely to be. And today, more and more companies, even those not explicitly in technology, need to master technology so they can move quickly enough to survive (Weill and Woerner in MIT Sloan Management Review 56(4):27, 2015 [1]). For new, digitally native businesses, that's no problem. They're built for speed. But more established businesses, even successful ones, have for many years delivered technology solutions to their employees and customers on lengthy release schedules that no longer make sense in today's accelerated environment. Based on our research and client work, we have developed and refined a two-speed architecture that lets more mature companies compete effectively with the upstarts.

Keywords Enterprise architecture · Digital · 2-speed architecture · Business enablement · Business transformation

8.1 Introduction

In recent years, established businesses have made many attempts to move faster. Some well-publicized approaches, like managing internal incubators and buying digital pure plays, have yielded mixed results. Transforming everything often serves as a sign of desperation. Our research is based on a structured review of enterprise architecture engagements across industries; an in-depth outside-in analysis of technology, governance, and organization of digital natives and startup companies; and work with our clients. Through our research, we wanted to learn how companies with a history could adapt successfully to the new digital world. Moving an

O. Bossert (✉)
McKinsey & Company, Frankfurt, Germany
e-mail: oliver_bossert@mckinsey.com

© Springer International Publishing Switzerland 2016   139
E. El-Sheikh et al. (eds.), *Emerging Trends in the Evolution of Service-Oriented and Enterprise Architectures*, Intelligent Systems Reference Library 111,
DOI 10.1007/978-3-319-40564-3_8

entire company to a fast speed all at once is unlikely to succeed, and that is true of a company's digital architecture as well. As we learned more, we discovered that the companies are more likely to succeed in going digital if they do so by adopting a two-speed approach [2].

The need for change is clear. To remain in existence, companies across a wide range of industries have adopted digital business models. Social networks and ecommerce websites have raised customer expectations regarding ease, speed, and agility. Consumers expect similar performance from their banks, retailers, and telecommunications companies. Many older companies are struggling to keep up with these expectations.

The ability to offer new products on a timely basis has become an important competitive differentiator. Some of the more advanced digital natives, such as Netflix, release new code within 1 h [3]. For larger companies at the forefront of digital, weekly or more frequent software releases for the digital platform have become common. Such speed is possible only through the inherently error-prone software-development approach of testing, failing, learning, adapting, and iterating rapidly. That experimental approach cannot be applied to legacy systems. Nor should it be, since key back-end legacy systems must be particularly resilient and stable. That is why many companies need to adopt an IT architecture that can operate at different speeds.

Unlike enterprises that are born digital, traditional companies do not have the luxury of starting with a clean slate. Established organizations must build an architecture designed for the digital enterprise with a legacy foundation. We identified the core elements that make such a transformation successful—and others more likely to lead to failure.

8.2   The Digital Era

The digital revolution not only changes our personal lives but it also has massive implications on the competitive landscape. To win in this new environment, established companies need to develop new products quickly, interact across channels, analyze customer behavior in real-time, and automate processes. Digital can lower barriers to entry for new players, causing long-understood boundaries between sectors to become more ambiguous and permeable. At the same time, the "plug and play" nature of digital assets disaggregates value chains, creating openings for focused, fast-moving competitors.

New market entrants often can scale rapidly at lower cost than legacy players, and returns may grow with astonishing speed as more customers join the new networks. Digital capabilities increasingly will determine which companies create or lose value. Those shifts take place in the context of industry evolution, which isn't monolithic but tend to follow a common path: new trends emerge and disruptive entrants appear, their products and services embraced by early adopters [4] (see Exhibit 8.1). Forward-looking incumbents then begin to adjust to these changes, accelerating the rate of customer adoption until the industry's level of digitization—among companies but more critically among consumers—reaches a tipping point. What was once radical becomes normal and unprepared incumbents run the risk of becoming disappearing as quickly and surely as Blockbuster. Other legacy players that successfully built new capabilities (as Burberry did in retailing), transform into powerful digital players.

Such a digital transformation requires companies to adapt faster to new trends and invest massively in digital. If organizations miss the tipping point, they often also miss the chance to further invest in technology. The agility of the company diminishes and it will eventually go out of business.

The digital trend impacts different industries differently. Our research shows that some industries will go through a radical reshaping transformation over the next years while others will be impacted more by what digital enables for them. As shown in Exhibit 8.2, customer-facing industries tend to be the ones most impacted by digital.

The digital revolution is driven by the end-customer—the consumer. As shown in Exhibit 8.3, most companies identify the digital engagement of consumers as their number one strategic priority among digital trends. Just 34 % of executives say it's a Top Three priority for their companies, even though automation may significantly help businesses in sectors that are undergoing digital disruption [5].

The power of customers become apparent when looking at the extraordinary growth rates of new offerings that directly address the end-customer—such as Uber in transportation and AirBnB in lodging. More than one-third of all C-level respondents said they expect at least 15 % of their companies' growth in the next three years will be driven by digital. Their top priority is to tap into new profit pools and create new business models, addressing customer segments that are not yet in the companies' strategic focus (Exhibit 8.4).

Constructing a new digital business model for an incumbent player necessitates building on existing strengths. But adding to existing capabilities also requires building atop the existing IT landscape. In contrast to "digital native" players, most large incumbents have an architecture built over time that acts not as a competitive weapon but as a hindrance to innovation. What is in those legacy systems can't just be abandoned; the data is crucial to the overall business's survival. For a retailer, the only possible way to compete with the pure-play online ecommerce stores is to establish a solid omnichannel strategy. This strategy requires transparency into store inventories managed by an outdated legacy system. It may be outdated, but it is still necessary.

At the same time, companies have to move quickly if they do not want to be disrupted. The speed of digital transformation doesn't allow for a year-long transformation but rather requires an architecture of two speeds. That fast speed will be different in different industries. For a public agency, fast may mean a new deployment every two weeks when it used to be every three months. For digitally native Netflix, which offers best-in-class continuous updates, it means as many as 100 deployments a day [6]. It's all relative to how quickly an organization's customers are expecting changes to the user experience of a certain company.

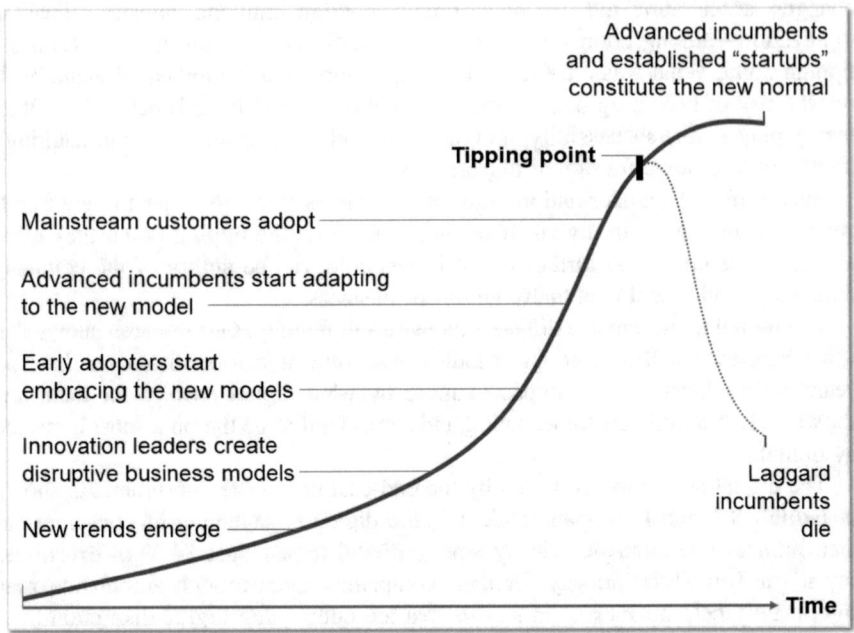

Exhibit 8.1 Schematic development of new digital trends and their impact on established business models

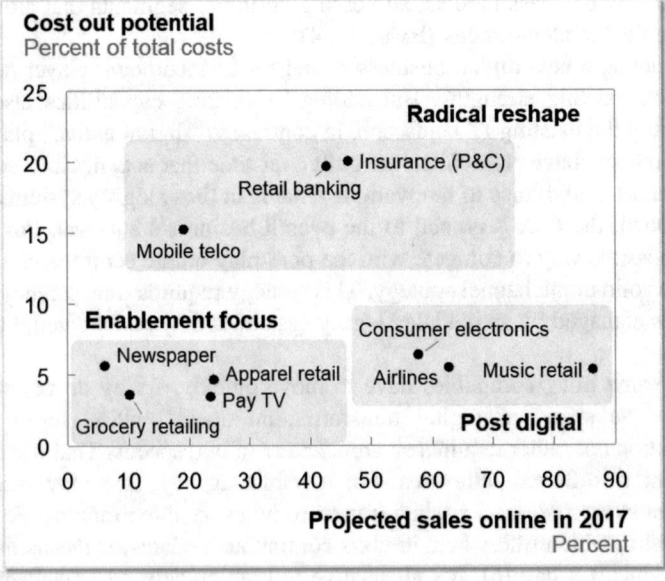

Exhibit 8.2 Analysis segmenting industries by their cost out potential and projected online sales

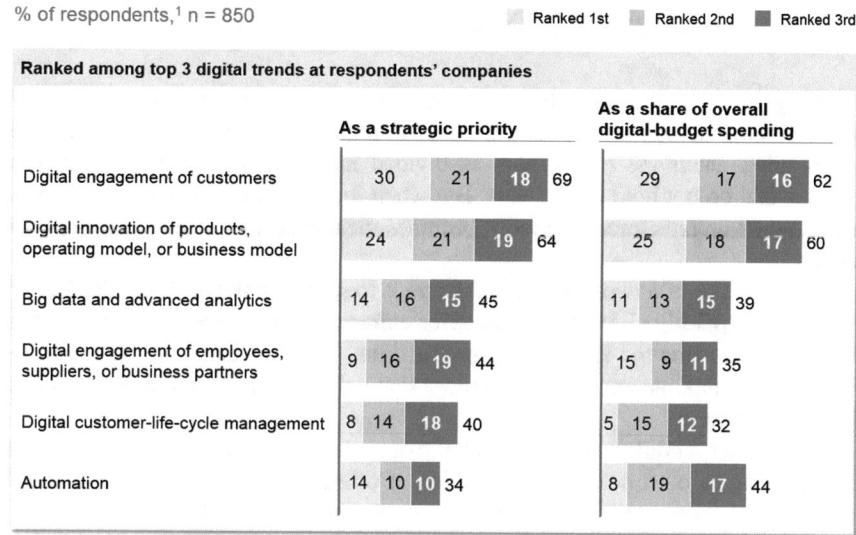

Exhibit 8.3 Results of a global survey among senior executions on their view on digital trends

Exhibit 8.4 Results from the same survey as above analyzing the key objects of digital growth

8.3 Fundamentals of a Two-Speed Architecture

Layering is not new to IT. It is easy to confuse a two-speed architecture with traditional technology layers, such as business logic and the presentation layer. One might think about the two speeds as divided into systems of record and other systems that do not hold master data. But when focusing on the desired outcome of a two-speed architecture, it quickly becomes clear that these views are not quite correct.

The objective of a two-speed architecture is to **separate those elements that are required to quickly change the customer experience** from other elements that are more important for the integrity of transactions. It differentiates the systems that must be most flexible and agile from those that have to be more reliable and deliver the highest quality.

Some systems containing business logic or master data records are pivotal for the implementation of this value proposition. Consider the fast architecture that a retailer has implemented (see Exhibit 8.5). A two-speed architecture has to cut across different layers of the technology stack:

- The fast-speed-architecture contains the channels that are pivotal for the customer experience.
- To change the customer experience, companies have to do more than alter the front end presentation layer. New functionalities (such as a new loyalty program) have to be implemented at the same speed within the same deployment cycle. Changing the design without changing what the design enables is not enough and merely delays the more important work.
- Some of the systems of record also have to move into the fast speed architecture. Customer data structures as well as product data have to change quickly to enable new business models. Without them, companies cannot enable a differentiating customer experience.

Digitally native businesses can excel at giving consumers what they want, while many older companies struggle to meet current customer expectations. For established businesses, success requires strong capabilities in four areas:

First, because the digital-business model lets businesses create digital product services and bring them to market quickly, **companies need to become skilled at digital product innovation that meet changed customer expectations**. Consider the new generation of auto insurance policies, enabled by geotracking technology, that determine the price of a premium based on how much and how aggressively an individual actually drives.

Second, **companies need to provide a seamless multichannel experience**, both in the digital and physical realms, so consumers can move effortlessly from one channel to another. Many shoppers use smartphones to reserve a product online and then pick it up in person at a store.

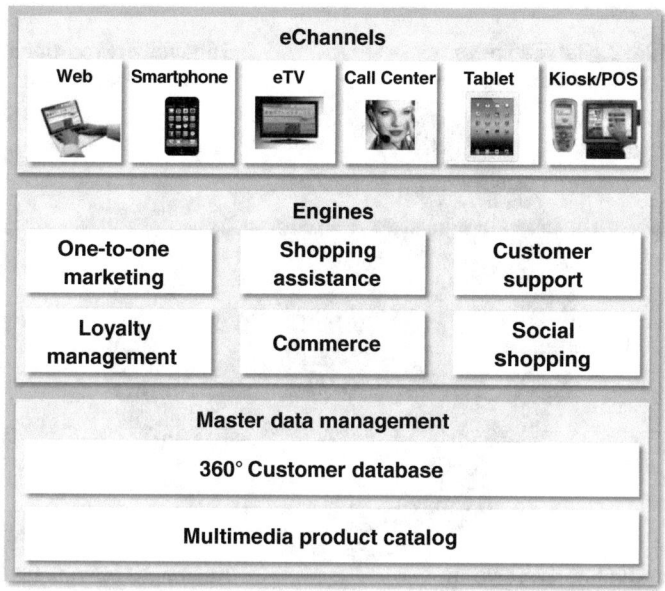

Exhibit 8.5 Retail architecture bundling the necessary digital capabilities in one fast speed platform

Third, **companies should use big data and advanced analytics to better understand customer behavior**. Gaining insight into customers' buying habits—with their consent, of course—can lead to improved customer experience and increased sales.

Fourth, **companies need to improve their capabilities in automating operations and digitizing business processes**. This is a sure way to enable quicker response times to customers while cutting operating costs (Exhibit 8.6).

8.3.1 Implications for Enterprise Architecture

Each of these four levers poses a substantial challenge for IT. For example, many banking-product lines—such as credit cards, investments, and checking and savings accounts—are managed in silos. This makes it difficult for a business to quickly get a comprehensive view of customers to assess their loan applications. What's more, channels are often managed and tracked independently, complicating matters for customers who wish to use multiple channels as they pursue a transaction. For example, customers who start a loan application on their smartphone often find that they have to re-enter data when they switch to a desktop computer to fill in the more

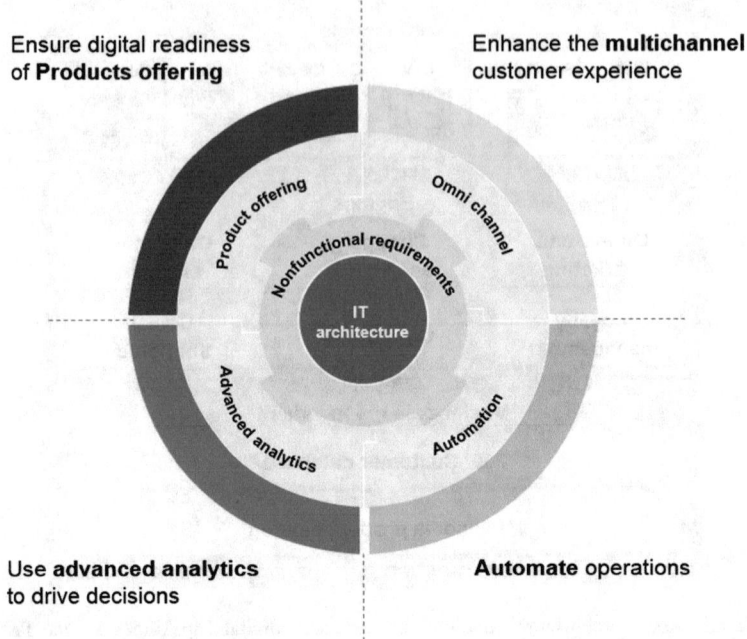

Exhibit 8.6 Core value levels of a 2-speed architecture

detailed information required. Weak systems integration and slow database-access times prevent customers from enjoying a real-time shopping and purchasing experience. Analytics is especially difficult to integrate with operational process flows. Manual steps in these processes, such as rekeying and transferring information, present major obstacles to both analytics and automation of processes. They also discourage potential and existing customers.

While a handful of players have overcome some of these hurdles, implementing all four levers so customers can purchase individually tailored products across multiple channels is quite a challenge. Legacy IT architectures and organizations that run the supply-chain and operations systems responsible for executing online product orders often lack the speed and flexibility needed to survive in the digital marketplace.

Offering new products in a timely manner has become a competitive differentiator. For an ecommerce platform, this might require weekly software releases. That kind of speed can only be achieved with an inherently error-prone software-development approach of testing, failing, learning, adapting, and iterating rapidly. That experimental approach would not work with legacy systems, where the demand for perfection is far higher. Quality comes slowly but is critical for risk- and regulatory-compliance management, as well as for core transactional activities such as finance and online sales. In contrast, lower IT-system quality and resilience can be acceptable in customer-facing areas. This is particularly true when users participate in the testing of new software. They expect to encounter mistakes.

8.4   The Building Blocks of Digital-Enterprise Architecture

In our experience, digital-enterprise architecture needs to accommodate the following elements to deliver the functionality that the digital enterprise requires (Exhibit 8.7).

Two-speed architecture: This implies a fast-speed, customer-centric part of the architecture running alongside a slower release cycle, transaction-focused stable architecture. For software-release cycles and deployment mechanisms, the customer-facing part should be modular. This permits quick deployment of new software by avoiding time-consuming integration work. In contrast, the transactional core systems must be designed for stability and high-quality data management, hence their longer release cycles.

Instant cross-channel deployment of functionality: New microservices defining only a small amount of functionality, such as lookup of the next product a consumer might purchase, should be deployed in hours rather than weeks. Such

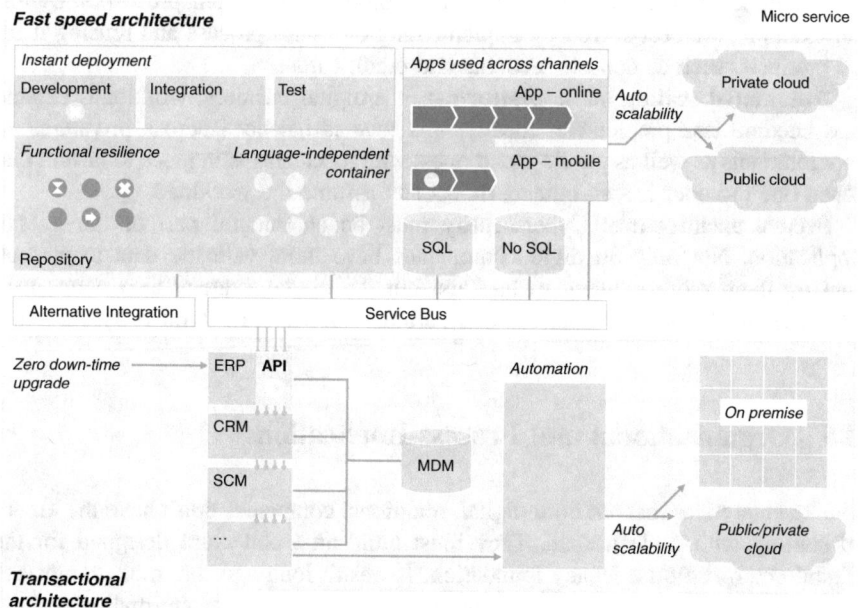

Exhibit 8.7 Reference architecture showing the building blocks of a 2-speed architecture

microservices should be available across all channels. Ideally, developers should create these services in multiple programming languages rather than being locked into a single development framework and should be able to deploy them into production following a DevOps approach.

Zero downtime: In digital global operations, the era of days-long maintenance windows is long gone. Upgrades of systems affecting the consumer's experience should be seamless. They are best achieved by deploying a new software or service in parallel with the old version. At first, only a small percentage of user traffic should be routed to the new version. Then the switchover progresses. Only when the new version meets a set of key performance indicators will all traffic be routed to the new version. Moreover, in daily operations there should be fallback mechanisms in place so that issues arising in one service do not harm overall operations. If, for example, a retailer's personalized recommendation service is unavailable, a random recommendation could be displayed rather than an annoying error page.

Real-time data analytics: Customers generate data with every move they make within an app. That makes analytics an integral part of operational processes. For example, one retailer analyzes customers' purchases automatically when they pay with their credit card. Along with the receipt, the business provides a savings coupon for a product customers may be interested in buying the next time they shop at the store.

Easy process configuration: Business users should be able to change automated processes without requiring time-consuming coding by an IT developer.

Product factory: Industries that provide digital products, such as banking and telecommunications, need to decouple their products from their processes. Banks, for example, are best-served by implementing one sales process and reusing it for all products, such as deposit accounts and credit cards.

Automated scaling of IT platforms: In a digital business, workloads expand and become less predictable. Ideally, this load is balanced across private-cloud environments as well as public-cloud ones, with mechanisms in place to ensure that when one provider has an outage, others can assume the workload.

Secure architecture: Cybersecurity must be an integral part of the overall application. Not only do digital companies have more valuable data to protect, making them more enticing to hackers, but the digital strategy also opens new interfaces to customers, suppliers, and partners that can be exploited by hackers.

8.5 Organizational and Process Implications

Unlike enterprises that are born digital, traditional companies don't have the luxury of starting with a clean slate. They must build an architecture designed for the digital enterprise on a legacy foundation. It wasn't long ago that most companies would have been comfortable enduring a three-to-five-year transformation and not implementing new features until the process was complete. Today's highly competitive markets no longer allow altering architecture and business models sequentially. Digital transformation is a continuous process of delivering new functionality. Successful digital transformations focus on the following aspects:

Design the target architecture: The transformation can only be sustained if a high-level target architecture and standards in critical areas such as cybersecurity are described clearly from the beginning. Without target architecture and a sound security concept in place, any transformation can be slowed down by the complexity of managing legacy and new hardware and application provisioning.

View the transformation as a continuous software-solution process: There isn't time to develop software by using a waterfall model and then separating the transformation into several long phases, as was common in traditional multiyear IT transformations. Now, the software solution for each business challenge has to be continuously developed, tested, and implemented in an integrated fashion.

Start the transformation with a pilot: The transformation should be based on agile software development and the advanced governance processes that come with it. For companies new to agile, we advise piloting the methodology and tailoring it to their specific needs in a smaller-scale context.

Develop the transactional architecture, too: It's important to establish a clear distinction between the two IT models from the beginning and not only focus on the fast-speed part but also develop the transactional back-end architecture. One should not be pursued at the expense of the other.

Build a new organization and governance model in parallel with the new technology: In the digital enterprise, business and IT work together in new and integrated ways. Boundaries blur. This partnership has to be established and solidified during the transformation.

Change mindsets: By transforming the architecture, technology can become a key factor for a company's competitiveness. This requires increased management attention and, usually, a place on the board agenda. While IT efficiency remains important, spending levels may rise as companies transform IT from necessary expense to true business enabler. As such, expenses must now be thought of as investments rather than just costs.

Run waves of change in three parallel streams: In a two-speed transformation, it makes sense to have an implementation plan that runs in three parallel streams. The *digital-transformation stream* builds new functionality for the business, supported by the results of a *short-term optimization stream*. That stream develops solutions that might not always be compliant with the target architecture. To ease the development of short-term measures and to create a sustainable IT infrastructure, an *architecture-transformation stream* is a third necessary component.

When an IT organization releases new digital functions on a faster deployment cycle, new levels of agility and coordination emerge, and they may require substantial organizational change. One large industrial company recently established digital product management as a separate organizational unit that owns the company's website, mobile applications, and digital interactions; has accountability for new functionality; and collaborates closely with business and IT leaders.

Creating joint IT-business teams to coordinate new initiatives proved invaluable at a bank trying to catch up with rivals. It used big data and advanced analytics to change products and marketing on the fly in response to evolving customer preferences. Product specialists now collaborate closely with model builders to create the automated tools that assess customer needs in real time and offer related financial products. The IT organization collaborates closely as it selects the best data processing technologies to support the new algorithmic models. None of this compromises the bank's transactional backbone, which is managed separately to ensure its ongoing integrity.

8.6   Conclusion

This was intended as a research endeavor focusing on how established companies could best manage digital transformations. As we developed the two-speed architecture and saw it in practice, we learned that the two-speed architecture was as much about organizational architecture and process architecture as it was about technology architecture. The practices we suggest in this paper don't merely spell out how a technology organization could thrive at two speeds, but how an entire enterprise can benefit from thinking that way. Best of all, we saw that companies adopting a two-speed architecture could build businesses out of their internal improvements, offering them as white-label solutions to other companies, even competitors. To create and sustain a successful digital strategy that may lead to unexpected benefits, businesses have to build up and stick to a two-speed strategy.

References

1. Weill, P., Woerner, S.L.: Thriving in an increasingly digital ecosystem. MIT Sloan Manage. Rev. **54**(4), 27 (Summer 2015 Research Feature, 16 June 2015)
2. Bossert, O., Ip, C., Laartz, J.: A Two-Speed IT Architecture for the Digital Enterprise. McKinsey on Business Technology (2014)
3. Bowles, G.: Self Service Build and Deployment at Netflix (Agile 2013). Engineering Tools at Netflix. http://de.slideshare.net/garethbowles/self-servicebuilddeploymentagile2013
4. Hirt, M., Willmott, P.: Strategic Principles for Competing in the Digital Age. McKinsey Quarterly (2014)
5. Willmott, P., Gottlieb, J.: The Digital Tipping Point: McKinsey Global Survey results. McKinsey Quarterly (2014)
6. Schmaus, B.: Deploying the Netflix API, 14 Aug 2013. http://techblog.netflix.com/2013/08/deploying-netflix-api.html

Chapter 9
Capability-Driven Development

A Novel Approach to Design Enterprise Capabilities

Hasan Koç, Jan-Christian Kuhr, Kurt Sandkuhl and Felix Timm

Abstract Technological advances, changes in regulations and increasing globalization of the economy demand high adaptability from enterprises in many areas. Enterprise Architecture Management provides organizations with an integrated view enabling such adaptability. In this respect, development and management of the capabilities receive attention, as the term is associated with flexibility, dynamics and variation. On the contrary, little effort has been put towards developing and modeling capabilities. This chapter focuses on the Capability-Driven Development (CDD) method, which is a novel approach for designing capabilities to tackle the challenges of rapidly changing enterprise environments by modeling the application context. The results presented in this chapter are (i) a description of the state of research in capability development methods, (ii) a component-wise structured capability modeling method based on business processes, goals and concepts of an enterprise, (iii) a demonstration of the method application in a use case from the utilities industry and (iv) observations made during the capability development and strategy use.

9.1 Introduction

Enterprises are confronted with rapidly changing situations in regulations, globalization, time-to-market pressures and advances in technology. In many industrial sectors, efficient and effective value creation and service delivery processes are considered as the key factors to competitiveness in a globalized market environment. Systematic management of enterprise architectures including the technical,

---

H. Koç (✉) · K. Sandkuhl · F. Timm
Chair of Business Information Systems, Institute of Computer Science,
The University of Rostock, Albert-Einstein-Str. 22, 18059 Rostock, Germany
e-mail: hasan.koc@uni-rostock.de

J.-C. Kuhr
SIV Software-Architektur & Technologie GmbH, Konrad-Zuse-Str. 1,
18184 Roggentin, Germany

© Springer International Publishing Switzerland 2016
E. El-Sheikh et al. (eds.), *Emerging Trends in the Evolution of Service-Oriented and Enterprise Architectures*, Intelligent Systems Reference Library 111,
DOI 10.1007/978-3-319-40564-3_9

application and business architecture is emerging into an important discipline in enterprises. One of the objectives of this discipline is to manage and systematically develop the capabilities of an enterprise, which often are reflected in the business services offered to customers and the technical services associated to them. In this context, networked enterprises, value networks and extended enterprises massively use service-oriented and process-oriented architectures.

The notion of capability has received a lot of attention as the enabler of business/IT alignment in changing environments. The term is used in various industrial and academic contexts with often different meanings. Most conceptualizations of the term agree that capability includes the *ability* to do something (know-how, organizational preparedness, appropriate competences) and the *capacity* for actual delivery in an application context. In Information Systems (IS) field the term capability is used as an instrument to develop Enterprise Architecture (EA) and tools for tackling the dynamic complexity of systems with diverse concerns. This indicates that flexibility, dynamics and variation are attributes associated with capability.

Even though the capability concept is an important element in service-oriented architectures and enterprise information systems, little effort has been put towards developing and modeling capabilities. Most notably, the state of the art research has shown that the notion of context is neglected, which might provide the required flexibility in changing situations. This paper focuses on the Capability-Driven Development (CDD) method, which is a novel approach for designing capabilities to tackle the challenges of rapidly changing enterprise environments by considering and modeling the application context. To be more concrete, the paper investigates the background, core concepts and most important features of the CDD method. The research questions investigated in this chapter can be summarized as follows:

1. What are the current problems in capability modeling for enterprises and what are the requirements for a potentially successful solution?
2. Which methods are available in the literature to design capabilities as enablers of business/IT alignment and to what extent do they fulfill the industrial requirements?
3. What are the benefits and drawbacks of the application of CDD in an industrial setting?

The research method is design science oriented and consists of a combination of three different methods:

- Addressing the *relevance cycle*, we conducted exploratory case studies for elaborating frame conditions and pathways for adaptation in industry as well as for validating the CDD. In particular, the study included three cases (multiple-case study based on a literal replication), whereas only one case is detailed in this chapter (focus of RQ3, Sects. 9.4 and 9.5).
- In *rigor cycle* we use the applicable knowledge in the literature by investigating frameworks, models and methods that might help in solving the problem and for grounding the work in state of research (focus of RQ2, Sect. 9.3)

- In the *design cycle*, we develop the artifact in line with the inputs from both cycles, observe how the developed artifact behaves in these scenarios and refine the artifact after the gathered feedback in the evaluation. Here, we apply an argumentative, deductive approach for developing the methodological basis for capability development (focus of RQ1, Sect. 9.2).

The remainder of the paper is structured as follows: Sect. 9.2 investigates the need for capability-driven development and the resulting requirements by considering two cases from practice, one of them from utility industries and one from e-government. Section 9.3 presents the theoretical background for our work that includes the notion of capabilities, current research in context modeling and in the field of capability design methods. Section 9.4 introduces the Capability-Driven Development method with a focus on the components of the methodology and the tools supporting its implementation. In order to demonstrate the practical use of CDD, Sect. 9.5 discusses a case from utility industries and initial experiences collected. Section 9.6 concludes this chapter by summarizing the work and presenting recommendations.

9.2 Problem Investigation: The Need for Capability-Driven Development

Before presenting the CDD method in detail, this chapter motivates the purpose of the CDD. Therefore, a problem investigation comprising the elaboration of industrial needs for designing capabilities has to be conducted. Further, requirements towards the CDD method have to be derived that guide the method's development. While Sects. 9.2.1 and 9.2.2 discuss two industrial settings regarding capability design, Sect. 9.2.3 then derives the requirements. Please note that the method has been developed based on the requirements derived from both use cases whereas the work at hand demonstrates the CDD approach only in utility industries use case due to place limitations.

9.2.1 Flexible Business Services in Utility Industries

The *SIV group* is a vertically integrated German enterprise that specifically serves the utility industry (cf. Fig. 9.1). It acts both as an independent software vendor (*ISV*) and as a business service provider (*BSP*). The group has a longstanding market presence in developing and selling the industry-specific ERP platform *kVASy*®. The platform is widely used by public utilities in Germany, especially for the commodities *electricity*, *natural gas*, *district heating* and *water*. Within the European Union the former two markets are strictly regulated, which in turn has led to increasingly complex business relationships and market rules. In Germany, market rules are subject to change, driven by the need to optimize existing processes and/or by altered legal framework. In the past, these changes have considerably impacted the way in which business operates within the market.

The SIV group has a vital interest in rendering its ERP product as well as offering flexible services adaptable to continuously changing markets. New opportunities, altered regulations and a growing competition lead to a demand for solutions that are delivering business value in ever changing context situations [1].

Given the constantly rising complexity of the market, public utilities increasingly consider outsourcing of their business processes to external service providers. The SIV group offers such services for customers running kVASy® by its subsidiary *SIV Utility Services GmbH*. Of particular relevance are business processes that deal with the *exchange of data* between market partners (market communication). Given the complex interrelationships of the market, exchanged data may easily get into conflict with other data, thereby initiating a clearing procedure.

This is the principal scenario for the SIV group's use case in the CaaS project. For any occurring exception the BSP acts as a clearing center with costly manual interaction. This further causes organizational efforts, such as the arrangement of BSP's human resources schedule. Contractual agreements between the BSP and its clients have to support a dynamic routing behavior in order to decide whether or not an individual clearing case should be routed from the customer to the BSP (cf. Fig. 9.1). The routing decision depends on various factors such as the backlog size of the client, message type that has thrown an exception or the type of the commodity. The awareness of such context-specific information would facilitate CDD in the business process outsourcing use case. In Sect. 9.5 a detailed description of the SIV use case and a demonstration of CDD method are provided.

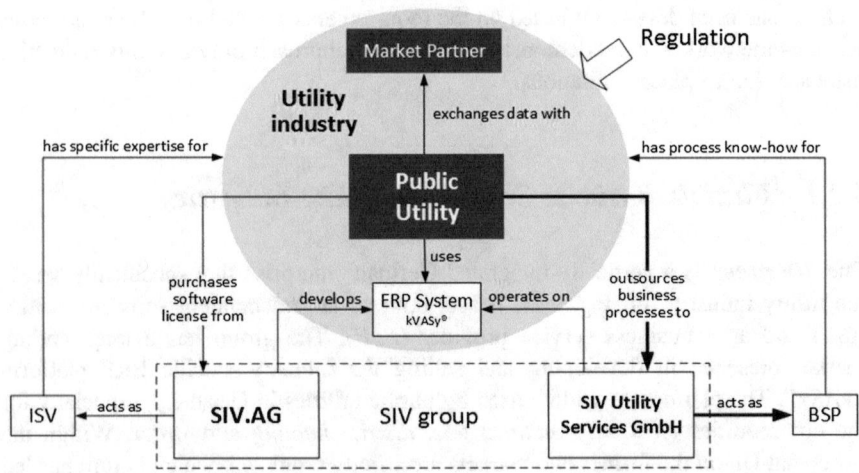

Fig. 9.1 Business relationships in the utility sector that are covered by the SIV group

9.2.2   Adaptive E-Government Services

everis is a multinational consulting firm providing business and strategy solutions, application development, maintenance, and outsourcing services. The everis use case is based on the public sector and the main emphasis is put on electronic services provided to municipalities, which are then used by citizens and companies. The company's service oriented architecture (SOA) platform provides a service catalogue with up to 100 services, which are active in 250 municipalities. Different factors and actors involved have to be taken into account when offering these services, such as small and medium-sized enterprises (SMEs), multinational corporations providing services, several public administration's laws, regulations, administrative consortia and calendars, as well as various technological tools. everis has to adapt the electronic services every time the platform is deployed to a new municipality and whenever the context changes. For the time being, service customization is done at the code level [2].

The SOA platform deployed at one municipality faces the challenge of irregular page visits and activity, which sometimes causes shortage of computational resources or unused resources. everis aims to optimize the utilization of computational resources. For instance, for a small number of daily visits of users the SOA platform could only use the necessary amount of computational resources. Identifying indications like social events would help to automatically integrate additional resources to the platform. This would ensure the fast and effective service execution in high load situations. Further, the integration of contextual data could be used to promote special services of the municipality enhancing the users' satisfaction with the platform [3].

A useful method for capability design considering all contextual influences should facilitate variability management and automate such customization as much as possible, flexible resource allocation as well as automatic service promotion.

9.2.3   Industrial Requirements

In order to serve as a basis for developing a CDD method, requirements were defined that the method should meet. Afterwards, current literature was analyzed regarding these prerequisites for already existing solutions for capability design the CDD could adapt. The results of this analysis are presented in Sect. 9.3. The CaaS project follows a use case-driven approach. Thus, requirements towards the CDD were derived from the industrial use cases presented above by conducting workshops and expert interviews with the industrial partners as well as analyzing secondary data. For each individual use case requirements were defined. Then, the resulting requirements set was merged into a set addressing the CDD approach as a whole, which also revealed synergies and differences among the use cases. Further, the authors of [4, 5] cross-examined the use case requirements based on industry-wide surveys that illustrated its relevance towards a wider user base rather than just the project's industrial partners.

The identified requirements are represented in a form of business goals. In [3] these goals were interrelated to each other with a goal model. As a result, 29 goals have been identified. This section concentrates on goals that are directly related to the use cases described in Sects. 9.2.1 and 9.2.2. The central goal is *to increase the value of the business services at hand* (REQ1), which is addressed towards the exception handling service of the SIV.AG as described in Sect. 9.5.2. It is directly supported by the goals related *to improve use case execution* (REQ2) that includes goals characterizing both capability design-time and run-time issues. While design-time addresses service customization according to its context and the identification of proper capability metrics, run-time considers adjusting the services by monitoring the context and change requirements prediction at delivery. Finally, to support different ways of working, the CDD has *to provide a flexible development method* (REQ3) by documenting the steps to design a capability, the important concepts as well as their representation.

During the requirements specification both enterprises raised the concern of dealing with variability and service line management. Thus, CDD has *to provide means for managing variability efficiently* (REQ4). As stated in earlier sections, *compliance with regulatory requirements* (REQ5) is a serious issue for SIV group operating in the utility industry and everis cooperating with governments. Further, it was revealed that both enterprises intend *to increase the level of process automation* (REQ6) due to service improvement and cost pressure. Another requirement is *to dynamically allocate computational and human resources for service delivery* (REQ7). This would help SIV group to meet SLA requirements as a BSP. Likewise, *the integration of internal and external information systems* (REQ8) was identified as a challenge for the CDD. For instance, everis needs the integration not only of municipality's information systems but also of sponsor systems and the usage of national services. While SIV group additionally identifies the need for the CDD *to develop new digital services* (REQ9), everis asks CDD *to increase the usage of services* (REQ10) delivered through the SOA platform by means of automatic promotion. In [4, 5] the process of developing the goals model is made transparent and illustrates the overall goals model with all goals explained.

9.3 Background and Related Work

The notion of capability has received a lot of attention as an instrument to align business and IT in changing environments to gain a competitive advantage. This section provides an overview of the related approaches to CDD. Section 9.3.1 introduces the notion of capabilities and positions our view within the Enterprise Architecture Management (EAM). Section 9.3.2 reports on context modeling approaches. Then, Sect. 9.3.3 presents the results of a systematic mapping study on the capability design methods and finally Sect. 9.3.4 summarizes the results.

9.3.1 Notion of Capability in CDD and EAM Capabilities

The term capability originates from the system engineering and military domain. Oxenham [6] uses it synonymously with *military capability*, meaning to apply the overall potential of the armed forces for combat or other operations. The literature analysis exposes three types of capabilities in IS field, such as organizational, system and operational capability. Boonpattarakan [7] considers organizational capabilities in the field of strategic management as the foundation in which organizations utilize their strengths to increase competitiveness, contribute to growth, and enhance organizational performance. The system capability describes the ability of a system to execute a particular course of action or achieve a desired effect, under a specified set of conditions. Finally, Adcock and Hoboken define the operational capability as the ability of a system to perform within the intended operational environment, particularly with respect to meeting the requirements of its stakeholders [8].

Although there seems to be an agreement about what constitutes a capability, it is hard to find a standard definition. Sandkuhl et al. [9] state that the definitions mainly put the focus on "combination of resources", "capacity to execute an activity", "perform better than competitors" and "possessed ability". A general consensus is that the capabilities are enablers of competitive advantage; they help companies to continuously deliver a certain business value in dynamically changing circumstances [10]. According to Chen and Tsou [11] the performance of an enterprise is the best, when the enterprise maps its capabilities to IT applications.

Enterprises are complex systems operating in changing environments. Ahlemann et al. [12] state that managing strategies, processes, applications, information infrastructures and roles is a challenge for an enterprise and there is a need to have an holistic view. Such an integrated view can be reached by implementing Enterprise Architecture Management (EAM). The IS domain adopts the term capability as an instrument to develop Enterprise Architecture (EA) and tools for tackling the dynamic complexity of systems with diverse concerns. For instance, Antunes et al. [13] identify the fostering of communication between stakeholders on technical and organizational levels as one of the most important merits of capabilities. Regarding Wißotzki et al. [14] an EAM capability describes the specific combination of know-how in terms of organizational knowledge, procedures and resources able to externalize this knowledge in a specific process with appropriate resources to achieve a specific outcome for a defined enterprise initiative.

We define capability as *the ability and capacity that enables an enterprise to achieve a business goal in a certain context* [3]. Ability refers to the level of available competence, where competence is understood as talent, intelligence and disposition, of a subject or enterprise to accomplish a goal; capacity means availability of resources, e.g., money, time, personnel, tools. This definition requires taking a slightly different view on capabilities as they are perceived in EAM. The first distinction lays in the investigation of operational aspects of an enterprise rather than management aspects, which necessarily challenges the integration of capabilities with enterprise modeling as well as their implementation (cf. REQ1, REQ2, REQ3 and REQ8). The modeling aspect includes analysis and design of enterprise models such as goals model, business process models or concept models. The implementation aspect consists of tools, transformation languages, notations and procedures. The second differentiation lies within the notion of context in the perception of what a capability is (cf. REQ4, REQ5, REQ6 and REQ7). As mentioned earlier, capabilities are useful instruments for enterprises in rapidly changing environments. Capabilities as such are directly related to the provision of business services and products, which are affected from the changes in the application context, such as, regulations, customer preferences and system performance. We hence argue that the notion of context plays a central role to support such flexibility, which is considered as *the adaptability to change* by Lacity et al. [15]. As a result capability application or delivery is closely related to the context of the business environment, which motivates the analysis of the notion of context and the context modeling approaches in the following section.

9.3.2 Context Modelling

Context is a widely used term in different areas such as philosophy, artificial intelligence, pragmatics, computational linguistics, computer science, and cognitive psychology. One of the most cited definition given by Dey and Abowd describes context as *"any information that can be used to characterize the situation of any entity"*, which is adopted in this work [16]. According to Winograd [17] this definition is too broad since *"something is context because of the way it is used in interpretation, not due to its inherent properties"*. Last but not least, Bazire and Brézillon identify main components of the concept "context" by examining a corpus of 150 definitions. The study concludes that context definitions can be analyzed in terms of six parameters like "constraint, influence, behavior, nature, structure and system". As a result, *"the context acts like a set of constraints that influence the behavior of a system (a user or a computer) embedded in a given task"* [18].

Although the term context is widely used in computer science, there is no general procedure how to develop context models. Many authors of context-based systems describe the way of developing the context model for their specific application, but do not provide a general view. Strang and Linnhoff-Popien provide a survey of six context modeling approaches in ubiquitous computing. These approaches consist of (i) key-value modeling, (ii) mark-up scheme modeling (*Comprehensive Structured Context Profiles, Pervasive Profile Description Language, ConteXtML,* etc.), (iii) graphical modeling (*UML, Object Role Modelling, ER,* etc.), (iv) object oriented modeling (*cues, Active Object Model*), (v) logic-based modeling and (vi) ontology-based modeling (*Context Ontology Language, CONtext Ontology,* etc.). The same article concludes that ontology-based modeling is the most suitable approach for context modeling for ubiquitous computing environments [19]. In this respect Gu et al. classify the existing context models into three categories. Application-oriented approaches are mostly used to represent low-level context information and lack formality. Model-oriented approaches utilize conceptual modeling techniques like ERM, UML and ORM. Ontology-oriented approaches intend to share knowledge across distributed systems. Finally, the authors present a context model based on ontology using OWL [20].

The state-of-the-art analysis conducted by Koç et al. [21] showed that context modeling and context-based systems are a popular topic in contemporary research with many different context definitions and application examples existing. Furthermore, the works mostly focus on the conceptualization of context, i.e., what elements context typically consists of and how to represent context models. An off-the-shelf context modeling method fulfilling the requirements and showing what steps to take as well as how to identify relevant context elements has not been proposed yet. However, the different context representations proposed can be used as inspiration based on the six parameters provided by Bazire and Brézillon [18], namely *constraint, influence, behavior, nature, structure,* and *system.*

9.3.3 Overview of Capability Design Methods

We conducted a systematic mapping study following the method of Kitchenham et al. [22]. To select the literature sources, we first identified A+ and A journals based on the rankings from Schrader and Hennig-Thurau [23]. Next, we complemented this list with B journals from the Business Information Systems sub-discipline. To stabilize the journal selection we crosschecked our results with the rankings from Peffers and Ya [24] and finalized the journal selection. After that we populated the list of journals with A and B ranked conferences from the work in [25], with "A" being the highest ranking. As a result, we identified a total of 112 journals and 24 conferences.

The main terms used for the initial search were "capability" (in abstract) and "method OR design OR proc*" (in keywords). The keyword terms were populated with additional terms "practice OR step OR modeling OR modelling". Consequently, we searched in the selected sources that included the term {capability} in abstract and one of the following terms {method, modeling, modelling, proc*, design, step, practice} in keyword. After removing the duplicates and inaccessible articles, the search resulted in a total of 362 journal articles and 178 conference papers in a time span from 1988 to 2014.

The selection of the papers was based on a set of criteria, which is applied during abstract reading. In cases where the exclusion or inclusion was unclear, an additional full-text reading is conducted. Articles exposing the following criteria were eliminated:

- articles that did not explicitly address "design and development of capabilities" as their research scope,

- articles using the term capability as a synonym for "ability" or "future" without relating it to its application in Information Systems,
- articles that use the term capability in the abstract and do not mention it in the narrative text or mention dynamic capabilities to position their proposals as means to gain competitive advantage,
- articles investigating the interrelation between capabilities and subject under study by applying statistical methods and developing a hypothesis.

The main purpose of the investigation was analyzing the state of the research in fields of capability design and development, in particular where processes, procedures, steps, or methods are proposed to identify, model or design business capabilities. To analyze the results, we used the method conceptualization of Goldkuhl et al. [26]. In a broader perspective, a method component consists of *concepts*, *activities* and a *notation*. The concepts specify what aspects of reality are regarded as relevant in the modeling process, i.e., what is important and what should be captured in a model. The activities describe in concrete terms how to identify the relevant concepts in a method component and the notation specifies how the result of the procedure should be documented. The articles are included that investigate steps, best practices, guidelines, concepts, notations and roles in developing capabilities. On the other hand, the publications were excluded that investigate how for instance the IT capabilities are interlinked with the provision of e- government services by applying statistical methods and developing hypothesis. After the application of those criteria, a total of 22 journal articles and 23 conference papers resulted for further analysis.[1] More details about the literature analysis can be found in the systematic mapping study of Koç [27].

One of the fields investigating the methods and approaches for designing capabilities is Business Process Management (BPM). Niehaves et al. analyze BPM topic from the Dynamic Capability point of view and present a framework to support the design of BPM capabilities. The framework developed in both works consists of three activities, namely sensing, seizing and transformation, which are further elaborated in sub-capabilities [28]. Ortbach et al. [29] offer a method for IT capability based business process design, which consists of 8 steps. The work describes roles and notations in vague terms, whereas the important concepts, outputs of the activities are not mentioned at all.

Another field investigating the capability design methods is the e-commerce. To exemplify a few works from our analysis, Cui and Pan present a process model to orchestrate the organizational resources in line with the changing business delivery context. However, no steps, activities and tasks are defined and the approach lacks notation to model the outputs of the phases [30]. Montealegre [31] suggests strategies for practitioners how to develop organizational capabilities in e-commerce and provides a process model including the key actions to be carried out.

---

[1]The list of investigated journals, conferences and selected papers can be accessed from http://bit.ly/1O5sS18.

In addition to the above mentioned topics, Zhou et al. [32] propose a process model for the development of capabilities to meet the fast growing business demands. Su [33] develops a theoretical model for conceptualizing the internationalization strategies of IT vendors, which consists of amongst others a capability-building process. Regarding both articles however, the processes cannot be applied solely as a capability design method since the roles, goals, concepts and notation is not defined. Besides, the latter contribution analyses the capabilities of IT service vendors in emerging economies, which remains too specific.

9.3.4 Summary

This section provided an overview of the related approaches to CDD. The results can be summarized as follows:

- CDD has a slight different view on capabilities than in EAM, primarily based on the following interpretations:

  - The CDD approach integrates organizational development with IS development by considering the application context and adjusting the solution in line with the dynamic changes. Thus, CDD assumes integration of capabilities with enterprise modeling, in particular with goals modeling and business process modeling.
  - In CDD, capabilities are directly related to the flexible provision of business services or products, which are affected from the changes in the application context. As a result, the notion of context plays a central role in design of capabilities.

- CDD uses a model-oriented approach to capture and represent contextual information. As we identified the lack of support for modeling context, CDD has to develop an own approach for context modeling including context representation and modeling process as such.
- There is a need to engineer a method since the approaches from the mapping study do not fulfill the requirements in general and REQ3 in particular. In this context, CDD provides a component-wise structured method to allow for flexible design of capabilities. The starting point for designing a capability can be enterprise goals, business processes or concepts, depending on the characteristics of an organization [34]. A central component to the method is the "Context Modelling Method Component".
- The capability design and development approaches provide to some extent support to design enterprise-grade capabilities, only partly fulfilling the requirements introduced in Sect. 9.2.3 as well as satisfying the method conceptualization concept.

To summarize, the analyzed approaches (i) address which factors should be taken into account when designing capabilities, (ii) show how they influence the capability under study (iii) provide means for capability evaluation, such as maturity models, (iv) do not address the integration of capabilities with enterprise modeling and finally (v) propose steps, procedures, guidelines to design capabilities that are decoupled from enterprise goals.

9.4   Capability-Driven Development

The industrial use cases discussed in Sect. 9.2 showed a need for supporting flexibility and adaptability in designing and developing IT-based business services. Furthermore, Sect. 9.3 analyzed existing work in areas related to capability management with the conclusion that the existing methodical and technological approaches in this area do not fully cover the needs discovered in the industrial use cases. This section proposes an approach to capability-driven design and development which aims at contributing to a novel paradigm for designing enterprise capabilities by explicitly taking into account business objectives and delivery contexts when developing IT-based business services. This novel approach is based on work in the EU-FP7 project "Capability as a Service in Digital Enterprises" (CaaS). From a technology perspective, the basic idea of the CaaS project is to facilitate a shift from the service-oriented paradigm to a capability delivery paradigm. Three industrial cases from utility industries, e-government and insurance and finance serve as basis development and validation of the approach for capability-driven development.

Business services are IT-based functionalities that digital enterprises provide for their customers. They usually serve specified business goals, are specified in a model-based way, and include service level definitions. To ease adaptation of business services to new delivery contexts, changes in customer processes or other legal environments, the CDD approach, is to explicitly define (a) the potential delivery context of a business service (i.e., all contexts in which the business service has to be potentially delivered), (b) the potential variants of the business service for the delivery context, and (c) what aspect of the delivery context would require what kind of variation or adaptation of the business service.

(a) The potential delivery context basically consists of a set of parameters or variables, the so-called context elements, which characterize the differences in delivery. The combination of all context elements and their possible ranges defines the context set, i.e., the problem space to cover.

(b) The potential variants of the business service, which form the solution space, are represented by process variants. Since in many delivery contexts it will be impractical to capture all possible variants, we propose to define patterns for the most frequent variants caused by context elements and to combine and instantiate these patterns to create actual solutions. If no suitable pattern is available, the conventional solution engineering process has to be used.

(c) The connection between context elements, patterns and business services has to be captured as transformation or mapping rules. These rules are defined during design time and interpreted during runtime.

The CDD method has to cover the development of all the above parts of a capability model and the CDD environment has to offer tool support for this purpose during design time and runtime. This section will describe both aspects of CDD in the following two sections.

9.4.1 CDD Method

The CaaS method for capability-driven design and development (CDD) follows a component-oriented method conceptualization with various components addressing different modeling aspects. The most important method components are as follows:

- Capability modeling supports identification and specification of capabilities and of relationships between capabilities. As a capability is characterized by business goals, business services implementing these goals, delivery contexts and patterns, a capability model basically integrates sub-models for these aspects. The capability modeling method component defines a procedure and aids for this integration.
- Enterprise modeling in general captures different perspectives of an organization (e.g., processes, organization structures, products, resources) and their interrelations in order to support purposes, like visualization of the existing situation, optimization of processes, business/IT alignment, or support of strategy development [35]. In CDD, enterprise modeling is used to capture strategic objectives or business goals related to the capability or motivating the creation of the capability. These objectives should be specified in a precise, measurable and accepted way, for example by using the goal modeling techniques from enterprise modeling.
- Business process modeling aims at specifying workflows and their execution. In CDD, the business service(s) offered to customers within the capability have to be specified using a model-based approach. Currently, the focus is on process-oriented approaches using process modeling.
- Context modeling: the potential application context(s) where the business service is supposed to be deployed has to be specified, which is supported by a newly developed context modeling method component. This specification also has to capture at what points in the process what variation will have to happen. The specification of the capability's potential deployment contexts is captured in a context model.

- During runtime, the execution of a business service in the defined context has to be monitored, including all variations of the solution for different context instances. As the context defines the switching between variants and potential parameterization of business services, a method component is needed to derive the configuration of monitoring application from context model and business service model.
- Pattern elicitation: patterns specify reusable design-time or run-time elements for reaching business goals under specific situational contexts. The run-time patterns are also called capability delivery patterns. The CDD method includes a method component for identification, elicitation and representation of patterns.

The relationships and interdependencies between different elements of the CDD method are illustrated in Fig. 9.2. The figure shows on the right hand side the *meta-model* and the modeling languages, which are core elements of the CDD approach (cf. [9, 36]). The meta-model defines all concepts and their interrelationships needed for patterns, contexts and enterprise models. The *CaaS modeling language* provides a visual modeling language for this meta-model, which is implemented in the capability design tool (cf. Sect. 9.4.2). In addition to the CaaS modeling language, the meta-model can also be implemented by other modeling languages, e.g., *established modeling languages* used in enterprises. The CaaS modeling language will be used primarily to model capabilities, contexts and patterns, as these models or views of the capability model are not available in enterprises before introducing the CDD approach. In contrast, enterprises already have enterprise models or models of their business services, which were developed with the modeling language established in the enterprise. If no modeling language is established, the CaaS language can be used for this purpose.

The middle part of Fig. 9.2 illustrates the different parts of a *capability model*, which also reflect the different method components introduced in the beginning of this section: a capability model includes the capabilities as such (*Capability View*), the context models defined for individual capabilities (*Context View*), the patterns used to implement adaptations in the capability during design time or runtime (*Pattern View*), the business goals related to the capability (part of the *Enterprise Model*) and the relation to the business service model (captured in the *Business Service Model*).

The left hand side of Fig. 9.2 shows method components related to runtime activities of CDD. The *Capability Delivery Navigation Application* (CNA) is an application which can be used to monitor and adjust the delivery of a capability during runtime (cf. Sect. 9.4.2). What has to be monitored in CNA is specified in the context model, i.e., the CNA has to be configured based on the context model. CDD provides a method element for the configuration process which implements the pre-defined *Configuration Options*. Furthermore, the runtime includes the executable IT-solution for providing the business service. However, this solution is not part of CDD.

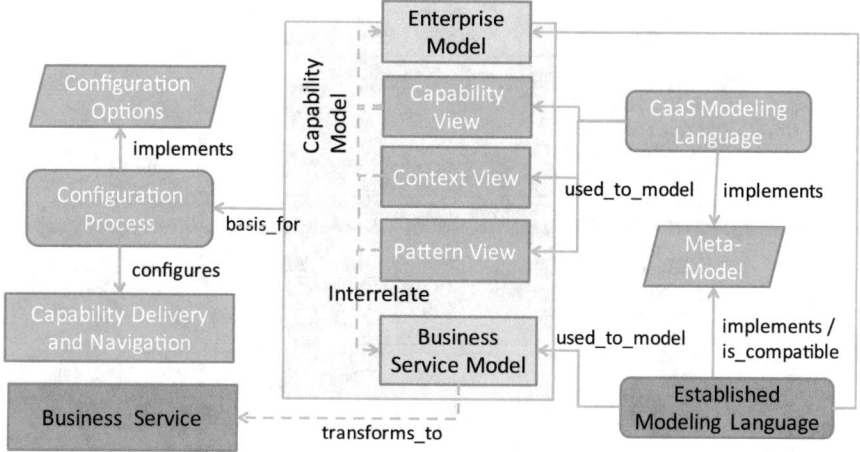

Fig. 9.2 CDD method components and their interrelations

The CDD method will be developed in several versions:

- CaaS base method: the main purpose of the initial CaaS method, also called "base method", is to support the industrial use cases of the CaaS project in developing initial capability models, i.e., the business services to be considered in the use cases including their context. For this purpose, the base method covers only selected ways of capability modeling and provide method components supporting these selected ways.
- CaaS method: the "regular" CaaS method will support a wider selection of capability development processes and extend the base method also towards capability delivery and runtime adaptation.
- CaaS method extensions: each of the industrial cases in CaaS are supposed to develop extensions of the regular CaaS method.
- Final CaaS method: one of the final results of the CaaS project will be a final version of the CDD method including the method extensions and packaged for use outside the CaaS project.

9.4.2   CDD Implementation

In order to implement the CDD approach, a capability development and delivery environment was designed and deployed in the industrial cases of the CaaS project. The architecture of the environment is shown in Fig. 9.3 and distinguishes components required at *runtime* for operating and monitoring capabilities, at *design time* for capability-driven development and as *resources* either for runtime or design time. Furthermore, Fig. 9.3 divides the architecture into functional components that have been developed in CaaS to implement the CDD approach (right side) and

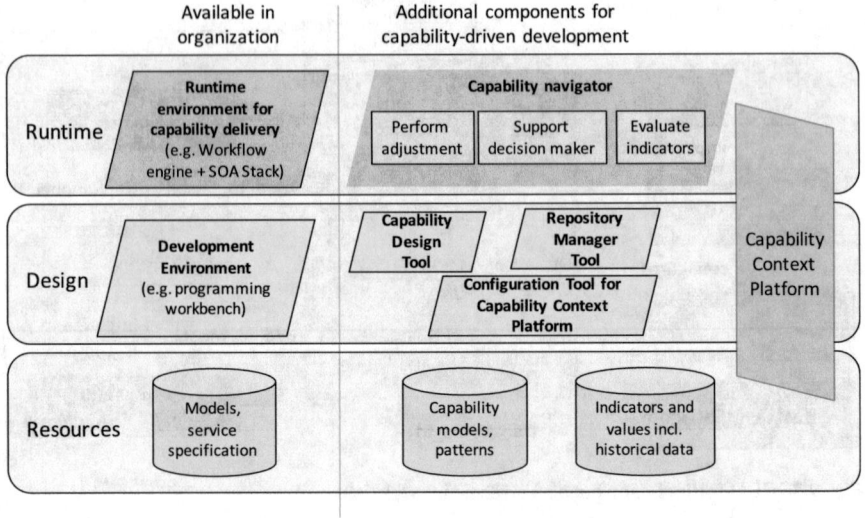

Fig. 9.3 Architecture and function components of the CDD environment

components that are expected to be available in organizations independently of CDD (left side). The latter include a *Development Environment* for IT-based business services (design time) and a *Runtime Environment* for delivery of the business services (runtime) and the capabilities based on these business services. The development environment could, for example, exist of a modeling environment for workflow models and a software development toolkit for developing services required when executing the workflows. The runtime environment—or Capability Delivery Application (CDA)—could consist of a workflow engine with activity monitor and an application server for the services integrated into the workflows.

The main functional components developed in CaaS for implementing the CDD approach are as follows:

- *Capability Design Tool* (design time)—provides modeling environment based on the CDD meta-model and CaaS modeling language, i.e., capabilities can be modeled including business service (e.g., business process model), business goals, context and relations to patterns.
- *Configuration Tool for Capability Context Platform* (design time)—the context model within the capability model specifies, which context elements have to be monitored during operations. The configuration tool for the capability context platform allows for connecting the design time view of context elements to the runtime view of data sources for these context elements, i.e., with this tool it can be specified which web service or database has to be used during runtime to get the actual values of context elements.
- *Repository Manager Tool* (design time)—manages creation, use and retrieval of patterns both for design time and for runtime use. The repository offers a defined

structure for storing patterns and a user interface for browsing the content. Furthermore, it offers interfaces which can be used by other design time and runtime tools.

- *Capability Context Platform* (CCP) (runtime)—captures data from external data sources including sensing hardware and Internet based services such as social networks. It aggregates data and provides these data to the capability navigator. It is configured using the configuration tool.
- *Capability Navigator* (runtime)—provides means for monitoring and adjustment of capability delivery. It includes a monitoring module for monitoring context and goal KPI, predictive evaluation of capability delivery performance and delivery adjustment algorithms. The capability delivery adjustment algorithms are built-in in the capability delivery navigation application. The algorithms continuously evaluate necessary adjustments and pass capability delivery adjustment commands to the capability delivery application.

Furthermore, there is an interface between the capability navigator and the runtime environment for capability delivery in order to be able to receive capability delivery adjustment commands from the capability delivery navigation application and to provide the capability delivery performance information.

9.5 Real-World Use Case: Utility Industry

9.5.1 Background and Motivation

In order to liberalize the utilities industry within the single European market, the European Commission since 1996 has enacted a number of directives that—in effect—make the unbundling of vertically integrated enterprises mandatory. In particular, grid operator and energy supplier are to be distinct market players. Non-discriminatory access to the grid must be granted to any interested party, such that there may be fair competition in the market.

In Germany, the Bundesnetzagentur[2] (BNetzA) is in charge of overseeing the utilities sector for the commodities electricity and natural gas and to ensure the European directives are put into national practice. To this end, the BNetzA has established a market role model that demands a clear separation of responsibilities among actors, of which an example is shown in Fig. 9.4.

While the consumer receives electricity from the *Balance Supplier* (step 1), it is the *Grid Access Provider* that takes care of the collection of the *meter readings*[3] (step 2). The latter forwards the meter readings periodically to the *Balance Supplier*

---

[2]cf. www.bundesnetzagentur.de.

[3]The actual reading is carried out by still another party (metered data collector), but this distinction is not relevant for the current use case.

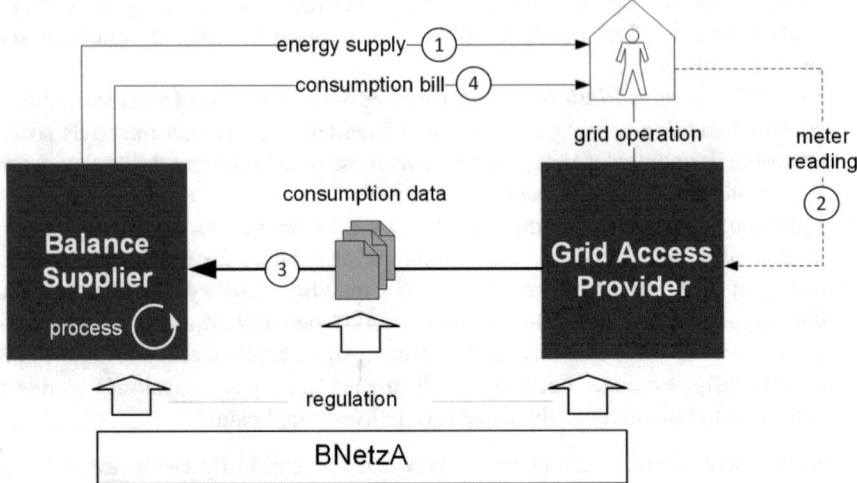

Fig. 9.4 Regulation of the energy distribution market in Germany makes the separation of market roles mandatory, where each of them performs distinct business functions. The underlying scenario is the electronic transmission of meter readings from a grid access provider to a balance supplier. These data are needed by the latter when invoicing the consumer with the energy consumption bill

(step 3) who invoices the customer on the basis of the transmitted *consumption data* by means of a *consumption bill* (step 4).

In this scenario, balance supplier and grid access provider perform distinct, yet complementary business functions. The interaction of both market roles is strictly bound to regulations enacted by the *BNetzA*. In Germany, data exchange between utilities and grid operators generally follows the EDIFACT standard, which is a widely used cross-industry specification for electronic business interactions. While the standard is issued and maintained by the United Nations, there exist specially adapted formats that regulatory authorities have made compulsory for the German market.

It is important to note that regulations are not fixed, but are rather subject to constant change. These changes can have significant impact on how companies run their businesses and as such they pose a constant challenge to utilities and ERP vendors likewise.

9.5.2 Use Case Scenario

This section details the use case that is currently developed and investigated by the SIV group to evaluate and demonstrate the CDD approach. The use case is based on the exchange of energy consumption data between grid access provider and supplier.

Such business-to-business interaction is a typical example of what is called *market communication*. Note that market partners usually run a cross-commodity business, including electricity, natural gas and district heating. In the utility sector, market communication is an indispensable and important value chain element of supplier and grid access provider likewise. Usually it requires the processing of bulk data that are transmitted within a single file.

For the exchange of energy consumption data, the BNetzA has mandated the use of the MSCONS format [2]. MSCONS is a member of the EDIFACT specification family and stands for *Metered Services Consumption*. German regulators have made the MSCONS format subject to biannual change, where each of them can impact the way how market players run their respective business processes.

Upon reception of an *MSCONS* file from a market partner, the *Balance Supplier* executes a business process to import the transmitted values into the ERP system (cf. Fig. 9.5). This process includes a file-level check, a validation step and the processing of the individual meter readings. Due to the complex nature of the market rules, meter readings are frequently found to be in conflict with other data such as master data. Unfortunately, many of these conflict situations cannot be resolved programmatically by the ERP system but rather require manual intervention by a domain expert (so-called clearing). The status of each of the client's business processes, including the ones that have failed, may be monitored by a business activity monitoring (*BAM*).

In an outsourcing scenario as shown in Fig. 9.5, clearing can be done either by the client or by the BSP. In the current use case, the *Business Service Provider* is *SIV Utility Services GmbH*, which is a member of the SIV group (cf. Fig. 9.1).

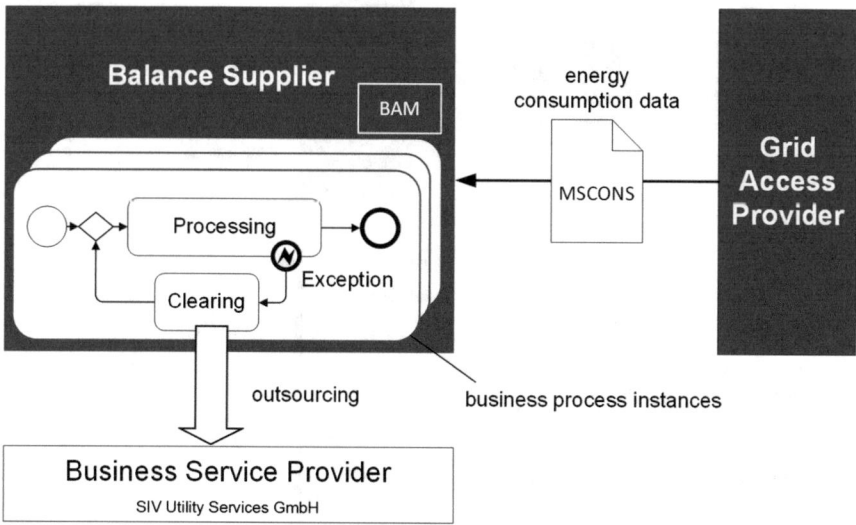

Fig. 9.5 Outsourcing of a supplier-run business process

A contract specifies operative conditions, such as commodity, type of measurement and the receivers' market role, upon which the service provision shall take care of failed MSCONS process instances. However, currently the service is completely manually operated and lacks a context-aware supporting system.

9.5.3 Capability Model

Balance supplier and grid access providers likewise expect from the BSP the capability to clear failed instances of market communication processes in a timely fashion and at low costs. In order to address this need with a CDD approach, a capability model has been developed, as shown in Fig. 9.6.

The model follows a goals first approach, i.e., capabilities are subordinate to an enterprise's vital goals. In the current use case, the goal of *Optimization of case throughput* drives the need for the capability *Clearing of failed instances*. Note that the SIV group's goals as illustrated in Fig. 9.6 are related to the use case and not necessarily derived from the requirements concerning the CDD method, as specified in Sect. 9.2.3. The implementation of this capability requires the execution of some business processes within the BSP's environment. Details on this process—which is typically modeled using the BPMN language—are shown in Fig. 9.8.

Context Model. In Fig. 9.6, the elements *Context Set*, *Context Element* and *Measurable Property* make those factors explicit that are relevant once the

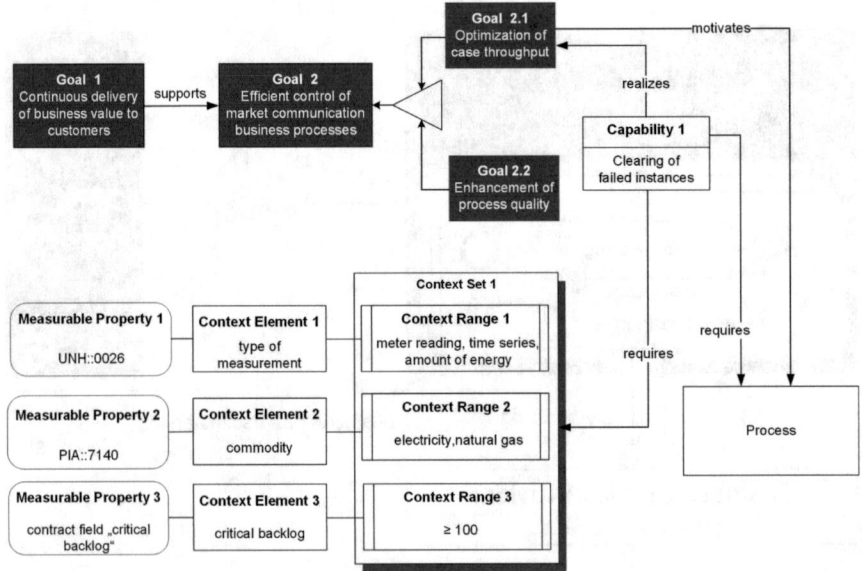

Fig. 9.6 Simplified capability model for the use case

capability in question is to be transferred to a different application context. An example of such an application context is a newly established market role.[4]

- A *Context Element* captures any characteristic information about a given entity [16].
- *Measurable Property* refers to any attribute that is to be measured in order to obtain the value of a context element.
- Capability deployment is usually subject to contextual restrictions. To capture this notion, context elements may be bound to *Context Ranges*.
- For each capability, all associated context ranges are grouped into a *Context Set*, such that each combination of context ranges constitutes an application context.

For the MSCONS scenario, important contextual factors are:

- the type of consumption measurement (e.g., the service may apply to meter readings only or to time series only),
- the commodity that is delivered to the consumer,
- the critical backlog for the client.

For each of these context elements exists a measurable property, which allows for the calculation of the context element's value. Such measurable properties may be obtained from the corresponding fields in the underlying contract, as illustrated in Fig. 9.6.

In a more general sense, even the message format itself may be considered a context element, i.e., the MSCONS clearing capability may well be extended to other EDIFACT formats, such as UTILMD and INVOIC.

Besides these factors, further context elements exist that are usually associated with the individual processing failure and determine the proper clearing procedure. All of these factors may be considered as *local context* since they refer to only a single client. We speak of *global context* if context data from all clients is aggregated to high-level key indicators. Such quantities can offer enhanced functionalities to the BSP, especially regarding the proactive allocation of human resources.

9.5.4 Clearing Center

The application of the CDD approach to the use scenario described in the last section has led to the notion of a *Clearing Center* (cf. Fig. 9.7). This subsection details the clearing center from a *conceptual* point of view.

The clearing center is an integrated application that comprises the components *CCP*, *CNA* and *CDA* (cf. Sect. 9.4.2). It is the technological platform that

---

[4]In 2011, German regulators have created the additional roles *metered data collector* and *meter operator*, which has increased the complexity of the utility market.

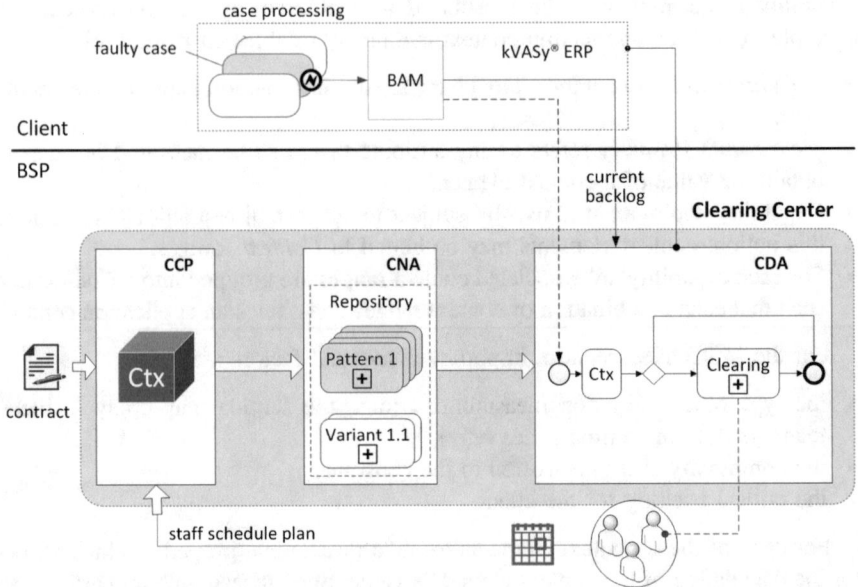

Fig. 9.7 Envisioned architecture of the *Clearing Center*

implements for the BSP the capability clearing of failed instances as described in the previous section (cf. Sect. 9.5.3; Fig. 9.6).

Upon the occurrence of a data processing failure in the client's environment, an exception is thrown. This event immediately triggers the creation of a corrective process instance in the BSP's system. Alongside with the exception, the relevant case data such as the conflicting file (MSCONS) is also forwarded to the BSP.

In order to clear the failed instances, there are a number of important *context elements* that need to be captured by the CCP and then evaluated and navigated by the CNA. These elements will help to significantly extend and enhance the as-is functionality of the BSP.

Contract. The *contract* between client and BSP is changeable but not volatile. It may specify quantities such as a list of commodities that are to be supported, a critical threshold of the client's backlog, and the client's market role.

Critical Backlog. The *backlog* tells the BSP the current number of outstanding failed instances at the client's side. In a simple scenario, action of the BSP may be required if the current backlog is greater than some contractually specified critical backlog. More realistically, this decision may also depend on other context factors, thereby offering greater flexibility and responsiveness to the client's demands.

Availability. The availability of BSP resources may also influence the runtime adjustments at the first variation point. This factor may be captured by binding an associated context element to the personnel deployment plan.

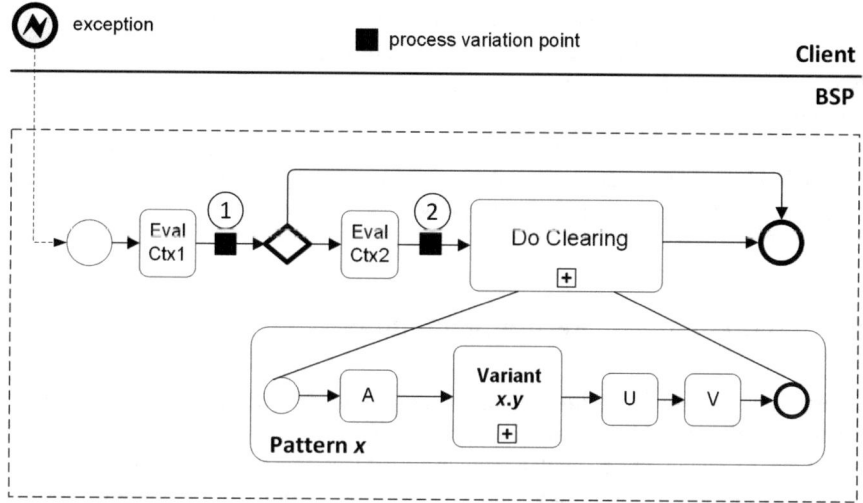

Fig. 9.8 Clearing business process as executed by the BSP

Figure 9.8 shows a business process that implements the clearing capability. It has two *variation points*, where each of them is driven by the context and depends on runtime adjustments done by the CNA.

Variation Point 1. The purpose of this variation point is to automatically determine whether action is to be taken by the BSP or not. This decision completely depends on the current context, which is accordingly evaluated by the task *Eval Ctx1*. Note that context may refer not only to the client's but also to the BSP's side.

The variation point has two possible outcomes, depending on whether the BSP is expected to do the clearing of the current case or not. In the latter event, the case is left with the client. However, statistical data may be collected by the CDA to support the client's future decision making.

In the former event, the BSP shall take action to clear the failed MSCONS process instance. As this decision is context-driven and the context may change over time, there can be no simple *built-in* rule that directs the control flow of the business process. Rather, the relevant information is *external* to the process. Again, to support the client's future decision making, statistical data are collected and made accessible for the client.

Variation Point 2. This variation point enables adjustments that substitute at runtime the actual clearing procedure by a proper business process (pattern). Patterns are recurring building blocks that may be used in many solutions. They may be considered best practices, so, in general, they can be anticipated at design time. A thorough analysis carried out to support the current use case has suggested that many of such patterns really exist and that they play a crucial role for delivering the BSP's capability to the clients.

Depending on the outcome of the task *Eval Ctx2*, the appropriate *Pattern x* is selected. However, the relevant context can only be determined at runtime. Navigation through the context requires access to data that are external to the current process instance, such as the contract between client and the BSP.

All of the pre-built patterns are registered in a CNA-held repository as shown in Fig. 9.7. Note that patterns may also contain process variants that correspond to minor deviations from a given best practice.

9.6 Summary and Recommendations

Enterprises operate in rapidly changing environments, which requires the implementation of adaptable solutions. This work introduced a capability management approach to tackle these challenges. The CDD approach consists primarily of a component-based methodology which at its core supports the modeling of application contexts (REQ4 and REQ5) and required adaptions of business services and tools supporting the CDD implementation at design time and runtime (REQ6, REQ7).

Experiences from evaluating capability management show so far that capability management is in particular promising if the following conditions are given:

- the business service under consideration exists in different variations and is subject to changes as soon as the application context changes,
- the business service is subject to the changes, e.g., by regulatory authorities or changing market environments, which cannot be planned, anticipated or controlled by the service provider,
- in order to optimize the service, different information sources have to be integrated during runtime.

When introducing CDD into an enterprise, the existing development methodologies and technologies do not have to be replaced or severely updated (REQ8). Capability management and CDD can be implemented as complementary approach. However, competences in CDD have to be developed within the enterprise.

Within the use case-driven CaaS project, three different strategies for capability modeling and design have been explored and elaborated for a flexible use in different industrial settings (REQ3). *Goals-first capability design* starts with the analysis of enterprise goals and defines how they can be reached in terms of capabilities, business processes and which context properties should be considered. Goals-first strategy is recommended for the enterprises that already have defined top-level organizational goals. The *process-first capability design* proposes that the starting point of the capability design is a process underlying a business service. The business service is further refined and extended by adding context awareness and adaptability, so as to establish a capability that can deliver this service in varying circumstances. Selecting this strategy requires the existence of detailed business process specifications or process-oriented culture. Last but not least, *concepts-first capability design* analyzes the existing knowledge structures in an enterprise and their relationships with the application context, which are essentially captured as concept models. For this, enterprises need pre-defined management structures, product structures or other conceptual models. A detailed comparison of the strategies can be found in [34].

The CDD use and "capability thinking" improves the understanding how business goals and technologies for implementing them are related. This contributes to the alignment of business and IT. In CDD, this is supported by the close relationship between enterprise model (includes goals, KPIs, business processes), capability design (includes explicitly defined application contexts) and capability delivery (runtime level for connecting the technological implementations in different application contexts). This kind of modeling can be used to analyze which IT-supported services would be suitable potentially for other customer groups or markets and what adaptations would have to be made for this purpose (REQ1).

By adopting the goals-first strategy in a use case from utilities industry, we developed a capability model to demonstrate the feasibility of the approach. The application of the CDD approach to the use case scenario described in Sect. 9.5 has led to the notion of a *clearing center*; a context-aware system offering flexible services responsive to customer demands (REQ2). Moreover, by modeling the goals and related business processes of the SIV group as well as making the application context explicit, it should be possible to develop new services in the continuously changing energy distribution market and/or increase the usage of existing services (REQ9). Experiences regarding the economic effects of CDD so far cannot be reported as the number of cases and the time frame available is not sufficient.

References

1. Goebel, C., Jacobsen, H., del Razo, V., et al.: Energy Informatics. Bus. Inf. Syst Eng **6**(1), 25–31 (2014). doi:10.1007/s12599-013-0304-2
2. España, S., González, T., Grabis, J., et al.: Capability-driven development of a SOA platform: a case study. In: Advanced Information Systems Engineering Workshops, LNBIP Vol. 178. Springer International Publishing, Cham, pp 100–111 (2014)
3. Bērziša, S., Bravos, G., Gonzalez, T., et al.: Capability driven development: an approach to designing digital enterprises. Bus. Inf. Syst. Eng. **57**(1), 15–25 (2015). doi:10.1007/s12599-014-0362-0
4. Bravos, G., Grabis, J., Henkel, M., et al.: Supporting evolving organizations: IS development methodology goals. In: Perspectives in Business Informatics Research, LNBIP, vol. 194, pp. 158–171. Springer International Publishing (2014)
5. Zdravkovic, J., Stirna, J., Kuhr, J., et al.: Requirements engineering for capability driven development. In: The Practice of Enterprise Modeling, LNBIP, vol. 197, pp. 193–207. Springer, Berlin (2014)
6. Oxenham, D.: The next great challenges in systems thinking: a defence perspective. Civil Eng. Environ. Syst. **27**(3), 231–241 (2010). doi:10.1080/10286608.2010.482661

7. Boonpattarakan, A.: Model of Thai small and medium sized enterprises' organizational capabilities: review and verification. JMR **4**(3), (2012). doi:10.5296/jmr.v4i3.1557
8. BKCASE Editorial Board: The guide to the systems engineering body of knowledge (SEBoK), v. 1.3. R.D. Adcock (EIC). The Trustees of the Stevens Institute of Technology, Hoboken, NJ (2014). Accessed 12 July 2015. www.sebokwiki.org. BKCASE is managed and maintained by the Stevens Institute of Technology Systems Engineering Research Center, the International Council on Systems Engineering, and the Institute of Electrical and Electronics Engineers Computer Society
9. Sandkuhl, K., Koç, H., Stirna, J.: Context-aware business services: technological support for business and IT-alignment. In: Business Information Systems Workshops, LNBIP, vol. 183, pp. 190–201. Springer International Publishing (2014)
10. Stirna, J., Grabis, J., Henkel, M., et al.: Capability driven development—an approach to support evolving organizations. In: The Practice of Enterprise Modeling, LNBIP, vol. 134, pp. 117–131. Springer, Berlin (2012)
11. Chen, J., Tsou, H.: Performance effects of 5IT6 capability, service process innovation, and the mediating role of customer service. J. Eng. Tech. Manage. **29**(1), 71–94 (2012). doi:10.1016/j.jengtecman.2011.09.007
12. Ahlemann, F., Stettiner, E., Messerschmidt, M., et al.: Strategic Enterprise Architecture Management. Springer, Berlin Heidelberg (2012)
13. Antunes, G., Barateiro. J., Becker, C., et al.: Modeling contextual concerns in enterprise architecture. In: 15th IEEE International Enterprise Distributed Object Computing Conference Workshops (EDOCW), pp. 3–10 (2011)
14. Wißotzki, M., Koç, H., Weichert, T., et al.: Development of an Enterprise Architecture Management Capability Catalog. In: Perspectives in Business Informatics Research, LNBIP, vol. 158, pp. 112–126. Springer, Berlin (2013)
15. Lacity, M.C., Khan, S.A., Willcocks, L.P.: A review of the IT outsourcing literature: insights for practice. J. Strateg. Inf. Syst. **18**(3), 130–146 (2009). doi:10.1016/j.jsis.2009.06.002
16. Dey, A.: Understanding and using context. Pers Ubiquitous Comput. **5**(1), 4–7 (2001). doi:10.1007/s007790170019
17. Winograd, T.: Architectures for Context. Human-Comp. Interact. **16**(2), 401–419 (2001). doi:10.1207/S15327051HCI16234_18
18. Bazire, M., Brézillon, P.: Understanding context before using it. In: Modeling and Using Context, LNCS, vol, 3554, pp. 29–40. Springer, Berlin (2005)
19. Strang, T., Linnhoff-Popien, C.: A context modeling survey. In: Workshop on Advanced Context Modelling, Reasoning and Management, UbiComp 2004—The Sixth International Conference on Ubiquitous Computing, Nottingham/England, pp 31–41 (2004)
20. Gu, T., Wang, X.H., Pung, H.K., et al.: An ontology-based context model in intelligent environments. In: Proceedings of Communication Networks and Distributed Systems Modelling and Simulation Conference, pp. 270–275 (2004)
21. Koç, H., Hennig, E., Jastram, S., et al.: State of the art in context modelling—a systematic literature review. In: Advanced Information Systems Engineering Workshops, LNBIP, vol. 178, pp. 53–64. Springer International Publishing, Cham (2014)
22. Kitchenham, B., Brereton, O.P., Budgen, D., et al.: Systematic literature reviews in software engineering—a systematic literature review. Inf. Softw. Technol. **51**(1), 7–15 (2009). doi:10.1016/j.infsof.2008.09.009
23. Schrader, U., Hennig-Thurau, T.: VHB-JOURQUAL2: method, results, and implications of the German academic association for business research's journal ranking, 14 Mar 2010
24. Peffers, K., Ya, T.: Identifying and evaluating the universe of outlets for information systems research: ranking the journals. J. Inf. Technol. Theory Appl. (JITTA) **5**(1) (2003)
25. WI: Die Sprecher der Wissenschaftlichen Kommission Wirtschaftsinformatik im Verband der Hochschullehrer für Betriebswirtschaft und des Fachbereichs Wirtschaftsinformatik der Gesellschaft für Informatik (GI-FB WI-Orientierungslisten. Wirtschaftsinformatik **50**(2), 155–163. doi:10.1365/s11576-008-0040-2

26. Goldkuhl, G., Lind, M., Seigerroth, U.: Method integration: the need for a learning perspective. Softw. IEE Proc. **145**(4), 113–118 (1998). doi:10.1049/ip-sen:19982197
27. Koç, H.: Methods in designing and developing capabilities: a systematic mapping study. In: The Practice of Enterprise Modelling, LNBIP, vol. 235, pp. 209–222. Springer International Publishing (2015)
28. Niehaves, B., Plattfaut, R., Sarker, S.: Understanding dynamic IS capabilities for effective process change: a theoretical framework and an empirical application. ICIS 2011 Proceedings (2011)
29. Ortbach, K., Plattfaut, R., Poppelbuß, J., et al.: A dynamic capability-based framework for business process management: theorizing and empirical application. In: 2012 45th Hawaii International Conference on System Sciences (HICSS), pp. 4287–4296 (2012)
30. Cui, M., Pan, S.L.: Developing focal capabilities for e-commerce adoption: a resource orchestration perspective. Inf. Manag. **52**(2), 200–209 (2015). doi:10.1016/j.im.2014.08.006
31. Montealegre, R.: A process model of capability development: lessons from the electronic commerce strategy at Bolsa de Valores de Guayaquil. Organ. Sci. **13**(5), 514–531 (2002). doi:10.1287/orsc.13.5.514.7808
32. Zhou, J., Zuo, M., Li, Q., et al.: Developing an agile it capability accompanying business's fast growing: a case study on a Chinese e-commerce company Ho Chi Minh City. In: PACIS 2012 Proceedings, Paper 24 (2012)
33. Su, N.: Internationalization strategies of IT vendors from emerging economies: the case of China. In: ICIS 2008 Proceedings. Paper 96 (2008)
34. España, S., Grabis, J., Henkel, M., et al.: Strategies for capability modelling: analysis based on initial experiences. In: Advanced Information Systems Engineering Workshops, LNBIP, vol. 215, pp. 40–52. Springer International Publishing, Cham (2015)
35. Sandkuhl, K., Stirna, J., Persson, A., et al.: Enterprise Modeling: Tackling Business Challenges with the 4EM Method. The Enterprise Engineering Series. Springer, Heidelberg (2014)
36. Zdravkovic, J., Stirna, J., Henkel, M., et al.: Modeling business capabilities and context dependent delivery by cloud services. In: Advanced Information Systems Engineering, LNCS, vol. 7908, pp. 369–383. Springer, Berlin (2013)

Chapter 10
Exploring the Nature of Capability Research

Matthias Wißotzki

Abstract Triggered by the progressive change from an industrial to an digital information society, not only social but also economic conditions are modified. Fast shifting business models and ever shorter product lifecycles are just few reasons why modern enterprises need a strategy how to deal with those unpredictable changes in order to stay competitive. Therefore, the concept of capability-driven management gets more and more attention by executives and scientists. In the last decade IS and management journals as well as conferences were publishing an increasing number of capability related articles, but a common understanding corresponding the identification of capabilities, their management, types or elements seems to be not existing. This work encapsulates the body of capability literature to provide an overview about capability research investigations over the last 15 years.

10.1 Introduction

Especially in times of virtualization enterprises are confronted with a lot of different challenges triggered by new technologies like big data, cloud computing, social business or cyber-physical systems. To handle these challenges, enterprises are using a composition of different management concepts. Especially, adaptability of IT focused business models or business and IT-alignment is supported by disciplines like Enterprise Architecture Management (EAM), Service Oriented Architecture (SOA) or IT Management (ITM). These disciplines expected to contribute to the above challenges by capturing the essential structures and processes of an enterprise on different architectural levels (e.g., business, data, application, technology), showing dependencies, supporting strategic planning and systematic development [41] (Fig. 10.1).

M. Wißotzki (✉)
University of Rostock, Rostock, Germany
e-mail: matthias.wissotzki@uni-rostock.de

© Springer International Publishing Switzerland 2016   179
E. El-Sheikh et al. (eds.), *Emerging Trends in the Evolution of Service-Oriented and Enterprise Architectures*, Intelligent Systems Reference Library 111,
DOI 10.1007/978-3-319-40564-3_10

However, enterprises have to reach certain goals through developing and implementing its strategies [40]. In order to reach these goals capability is a known term in science and practice. Capabilities appear in quite all functional areas of an enterprise from Business Process Management over Knowledge Management up to IT Management. Thus, understanding and methodology varies considerably from one approach to another which represents a challenge for companies to start in depth with capability driven management. Consequently, there is still an analytical lack of understanding and methods, regarding the commonalities and differences between different capability types and their research topics [14, 26, 35, 41].

What are capabilities of an enterprise and how they put together? Since there are plenty of books, journals and conference proceedings dealing with capabilities. This article deals with an extensive qualitative analysis within the capability research area intending a current state of research investigations. In order to figure out which research topics different capability types are referred to and to find similarities or differences between commonly used capability approaches, we used a systematic literature review approach executed in a larger exploration. Therefore, 21 persons in six teams from the Universities of Reutlingen and Rostock performed same structured SLAs based on the approach defined by Kitchenham [20].

10.2 Research Approach

This review should provide an overview about the capability research investigation in terms of abilities within enterprise context (not human or individual context) during the past 15 years. Therefore, we used a systematic literature review that supports a systematically identification, evaluation and interpretation of relevant sources in order to answer defined research questions by using a standardized process [20, 30]. Referred to [20] we performed three key stages and corresponding sub-steps that need to be processed to conduct a SLA. The first stage deals with the review planning and provides research questions (RQ), literature resources and time frame definitions for the investigation (Sect. 10.3). The second stage is called performing the review (Sect. 10.4), here we selected relevant articles and collect data for answering the RQs, realized in the final step: review report (Sect. 10.5), our conclusion and outlook is presented in Sect. 10.6.

10.3 Planning the Review

Considering the purposes of this work to be transparent, conclusive and give a comprehensive latest state, research questions had to deal with topics like research activity, research approaches, and statistics of identified concepts. All six teams received the same precisely described SLA process, same set of RQs and time frame. Just the literature source and databases were allocated in a team specific

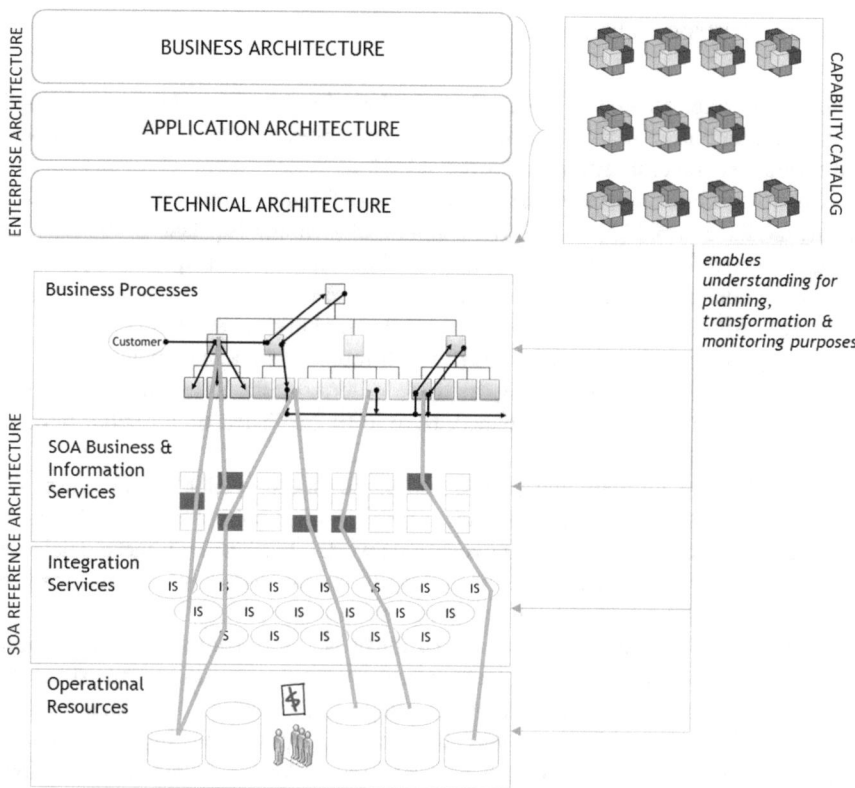

Fig. 10.1 Relationships between capabilities and EA (according to William and Rosen [39])

manner. The first review was executed in November/December 2014. Due to the fact that not all reviewed conferences and journals submitted their articles at the end of 2014, the author performed a second review just for 2014 in July 2015 by following exactly the same SLR process.

10.3.1 Motivating Research Questions

The first stage our investigation starts with the definition of research questions (RQ):

1. How has capability research been conducted within the last 15 years?
2. What research subjects are being investigated?
3. Who is active in this research area?
4. What research approaches are being used?
5. What capability definitions and descriptive elements are being used?

These RQs are answered based on found articles classified as relevant in Sect. 10.4.

10.3.2 Source Selection

After identifying research questions the selection of right sources has to be defined. With the intention to work on journals and conferences with high scientific impact, we formulated several criterions for the selection.

The first criterion relates to the journal and conference ranking that we assigned by the CORE Journal [3] and CORE Conference Ranking [9]. Due to the fact that not all selected journals and conferences rankings are provided by CORE [3, 9] we have been constrained to include another ranking provider like the Journal Quality List [18]. Moreover, in order to support the ranking criterion we include the H-Index (calculated by SCImago [32]) which represents another indicator for scientific impact as well. According to these criterions, the selected journals and conference have to be well established, have to be published on a regular basis in order to cover recent research topics and trends.

The published journals and papers had to be freely available. Therefore, the selected journals and conferences had to provide their publications on databases (DB) that are accessible by both university networks. Furthermore, these databases have to support the formulation of user-defined search-strings, which ensures that the reviewer teams could use unified terms and search rules.

Table 10.1 illustrates just an extract of thirty literature sources that fulfil these criteria The first column includes the literature source, the second informs about the utilized databases and the third about the impact. For instance, we analyzed the whole AISeL basket of journals and conferences supplemented by a selection of impact and content relevant journals and conferences known former investigations [42], like the Journal of Management, Information Systems Journal, The Practice of Enterprise Modeling, Perspectives in Business Informatics Research or IEEE International Conference on Commerce and Enterprise Computing.

10.3.3 Time Frame Selection

Identifying an appropriate time frame is considered to be the second step and reflects a contemporary state of research. Nevertheless, which period of time seems to be an appropriate one? For example, Simon et al. [34] selected a time frame of 23 years, McLean and DeLone [25] reviewed quite a half (1992 to mid-2002) whereas Urbach et al. reduced the period again by selecting articles published between 2003 and 2007 [30]. In the light of the above and in consideration of the journal and conference lifecycles we have chosen the period from 2000 to 2014. Thus, we searched for articles dealing with capability related topics published in defined literature sources within the last 15 years.

Table 10.1 Selection of literature sources and databases

Literature source description	DB	Rank/H-index
ISJ—Information systems journal	[18]	A [8, 17] H-Index: 52 [32]
SoSyM—Journal on software and systems modeling	[43]	B [8] H-Index: 28 [32]
JoM—Journal of management	[16]	A [17] H-Index 114 [32]
SMJ—Strategic management journal	[18]	A [17] H-Index 166 [32]
MISQ—Management information systems quarterly	[1, 10]	A [17] H-Index 132 [32]
JAIS—Journal of the association of information systems	[1]	A [8] H-Index 31 [32]
CAIS—Communications of the association for information systems	[1]	A [8] H-Index 15 [32]
BISE—Business and information systems engineering	[1]	n/a H-Index 18 [32]
SJIS—Scandinavian journal of information systems	[1]	A [8] H-Index 2
CAiSE—International conference on advanced information systems engineering	[43]	A [9]
ECIS—European conference on information systems	[1]	A [9]
HICSS—Hawaii international conference on system sciences	[15]	A [9]

10.4   Performing the Review

This phase includes selection of articles, relevance evaluation, data extraction, and data synthesis tasks. On the basis of that stage, we figured out which articles should be included in the final data extraction and synthesis regarding answering defined research questions.

10.4.1   Article Selection

In order to choose relevant articles from journals and conferences, search terms must be figured out and corresponding search strings has to be defined in the same way for all teams. Search terms should cover all possible keywords that are used to gather content-related articles. Moreover, in order to achieve an adequate result it was important to consider possible abbreviations and synonyms of the origin search term. The term 'Capability' is a quite strong term itself, because a common abbreviation does not exist [22, 27]. According to Refs. [22, 27] competence, skill, ability, aptitude are common synonyms, but they are commonly used in context of persons and individuals and not that relevant to describe the concept in terms of the defined research field (Sect. 10.2). Therefore, the following basis search terms were defined: capability and capabilities. In order to get articles primary dealing with a capability topic, it is assumed that the search terms are included in title or abstract. In particular, to answer RQ five we added secondary search terms to our search string that are optionally related to basis ones. Thus, the following conceptual search string for all reviews was defined:

> ((*"Capability"* OR *"Capabilities"*) AND (*"Identification"* OR *"Assessment"* OR *"Evaluation"* OR *"Framework"* OR *"Engineering"* OR *"Development"* OR *"Maturity"* OR *"Definition"*))

In this context, conceptual means that the search string syntax has to be slightly adjusted to the specific advanced search features of used databases (Table 10.1). Under consideration of the defined time frame and the titles and abstracts search all six teams performed the same relevance check procedure in order to eliminate the articles of the whole set of search results, that do not refer to the following restrictions: (1) The article has been deemed relevant by reading the abstract. (2) A search for primary and secondary search terms within the article was performed and the content found was classified as relevant. (3) If in doubt about the classification, the article was flagged for a second review (possible relevant). In order to increase the quality of the article selection process, different team members have performed the relevance classification of single articles. In order to evaluate article relevance we defined three categories: irrelevant, possible relevant and relevant. Articles classified as possible relevant have always been reviewed by second team member (second control). All classified as relevant articles were finally check by the project lead (third control).

Finally, after eliminating non-relevant articles and a third relevance check by the author we identified a total number of 190 relevant articles for answering our RQs. 144/191 articles were published in 18 conference proceedings and 46/190 articles in 12 journals. More than thirty countries and over 400 different authors were involved.

10.4.2   Data Collection

This section describes the data collection activities. Therefore, relevant classified papers were completely read, analyzed, data extracted and documented under consideration of the following aspects: journal/conference name, title, publication year, focus, research topic and method, authors and affiliation, capability definitions and descriptive/context elements, methods/frameworks/processes affiliation. We used these aspects as columns for our literature database and stored found information for each article, which provides the base for our review report.

10.5   Review Report

In order to answer the defined RQs this section presents and illustrates the findings and interpretations of the extracted data.

RQ1: How has capability research been conducted within the last 15 years?

In general, capability research has become more and more popular exemplified by an increasing trend of publications since 2000 (dotted line in Fig. 10.2). The first research activity we identified covered a resource-based perspective on IT capabilities and was published by the MISQ in 2000 [7], followed by an ECIS conference article regarding a theory of architectural knowledge integration capability in 2001 [36]. The first two peaks could be identified within half of the whole time frame. In 2005 and 2006 we counted 12 publications each with a 2/3 conference distribution. Nevertheless, from a journal perspective especially the MISQ published three of four noticeable articles regarding business and dynamic capability topics in 2006. Most intensive activities we recognized between 2009 and 2014 with over 20 articles/year in average with a majority of conference publications (76 %). Furthermore, it we documented a huge raise of published articles in 2011 compared to 2010. In this year the number of conference articles raises from 8 to 17 whereas the number of journal articles increases just by one. We locate an increasing interest in IT capabilities by researchers and practitioners, because more than half of the articles (10) were focused on that capability type. In 2008 we recognized a trend slump by more than 50 % from average 12.3 article/year down

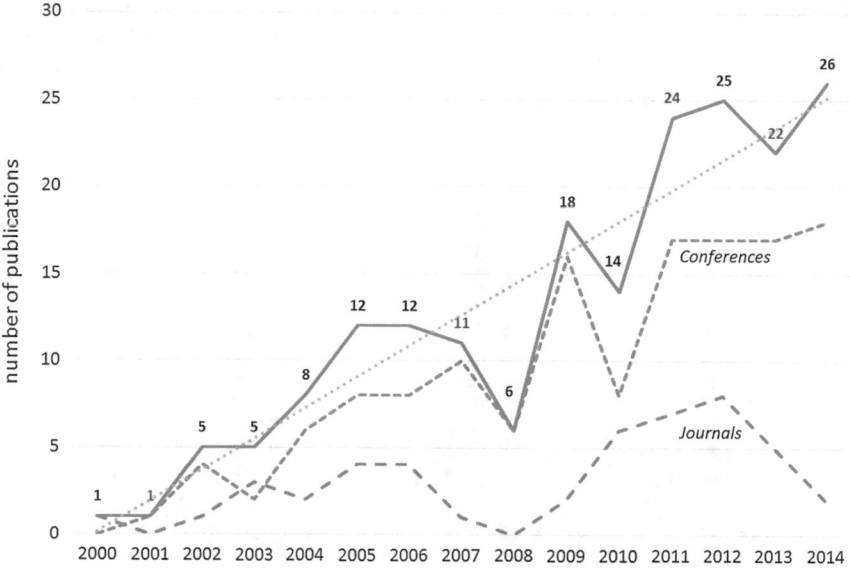

Fig. 10.2 Trend and number or relevant publications per year

to 6 papers. One reason for such an abrupt drift could be that the strongest publisher failed their average e.g., conferences like HICCS (0/2.1) with an average of 2.1 article/year published no article in 2008 or ECIS (1/1.8) with just one article. Journals like MISQ (0/1.34) and CAIS (0/0.54) published no articles as well in 2008.

For 2014, it cannot exclude that the number of 20 publications (found in the first run) could continue to rise, because not all analyzed conferences and journals already submitted their publications to the analyzed databases by the end of the year 2014. Therefore, we requested the literature sources (same procedure like in the first run) again, especially focused on 2014 (July 2015) and we found additional relevant 7 articles not published in December 2014 (5 PACIS, 1 MCIS). Finally the total number of relevant articles increases to 26 for 2014.

Not the amount of published articles per year but also the amount of articles published by each single journal or conference can be illustrated. To improve readability, Fig. 10.3 just illustrates journals and conference with more than one relevant publication at all.

The vast majority (32/191 articles) of article publications is attributable to the HICSS (17 %), followed by ECIS (15 %) and AMCIS (10 %). The MISQ (9 %) tops the list of journal article, followed by the JAIS (4 %) and the CAIS (4 %). Five journals ISJ, S and SMJ, SJIS, JGIM, IJKM, BISE published only one article in recent years and should be considered optional for future studies. Nevertheless, lifecycle information of journals and conferences should be considered in order to avoid premature decision regarding the thematic importance. For instance, BISE starts publishing in English since 2009, but the German version starts publishing in 1959 with changing names over the last decades and already enjoyed a wide standing in IS research (e.g., 1990–2008: Wirtschaftsinformatik). Thus, the number of article per year, and lifecycle information of a literature source combined with its impact (e.g., H-Index, Ranking, Impact Factor) are important aspects for the argumentation in terms of answering research questions.

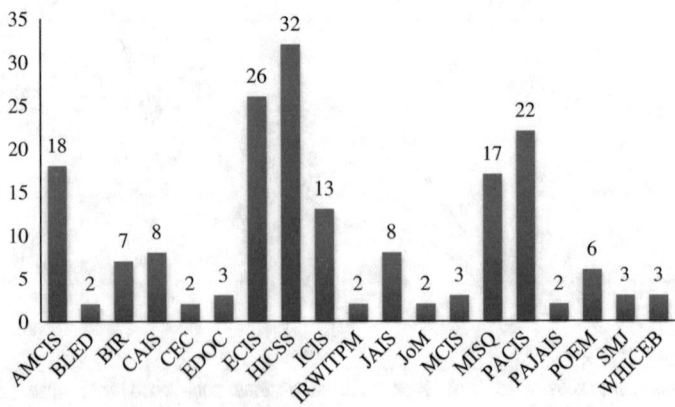

Fig. 10.3 Articles per journal and conference per year >1

RQ2: What research subjects are being investigated?

In the field of capabilities research the diversity of research investigations are increasingly widespread. In order to give an overview, we categorize all relevant articles and assign them to the following eight subjects:

(1) *Business Strategy Management*: contains all articles regarding strategic issues of a company like Process Change Management, Enterprise Transformation, organizational change, dynamic capabilities or the alignment of a company on E-Business, with positive effect on the business outcomes. (2) *Knowledge Management*: includes articles in the topic of Knowledge Management (e.g., Knowledge Transfer, Knowledge Integration, Data-Warehouse). (3) *Software Development*: covers articles that explicitly handle Software Engineering and Development or refer to the Capability Maturity Model. (4) *Project Management*: contains articles with an explicit referral to the topic of Project Management. (5) *Architecture Management*: covers articles with a holistic view on enterprise organization and architecture. Also includes elements of business process management, corporate performance management and alliance performance. (6) *IT-Management*: includes articles within the field of Information Technology- and Information System Management. This category contains the development, implementation and measurement of IT Systems as well as their impact on other categories. (NOTE: this category does not contain papers applied to the Software Development category). (7) *Supplier and Contract Management*: covers articles regarding suppliers or Supply Chain Management, as well as Contract Management and Sourcing Strategies. (8) *Development and Assessment processes*: contains articles within the topic of Business Process Management focused on development and assessment/measurement processes.

Several articles refer to more than one of the listed subjects or describe the impact of one subject to another. Therefore, these articles were assigned to more than one subject category. Figure 10.4 shows the number of articles assigned to their research subjects.

The reviewed scientific literature does not contain explicit articles about capability management, but we identified some articles regarding the nature of

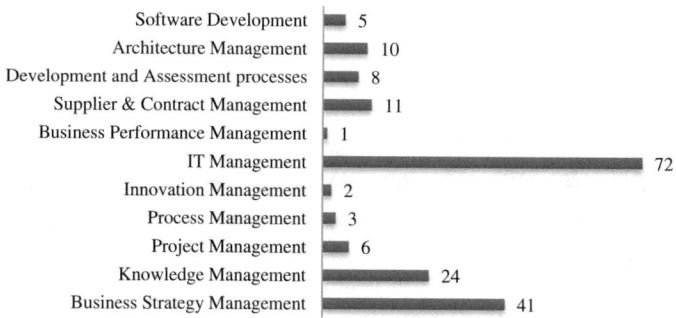

Fig. 10.4 Number of articles assigned to research subject

capabilities and capability modeling. Aside from software development and project management all subjects were discussed by more than one paper. Especially IT-Management (69) and Business Strategy Management (41) are focused topics within capability investigations. Knowledge Management (24), Supplier and Contract Management (10), Software Development (4), Project Management (6) and Architecture Management (10) would play a minor role. We recognized a fluent shift between Business Strategy Management and IT-Management, because former ones uses more and more IT-Management strategies and its capabilities for e.g., implementing business model, digitalize supply chains or communication. They represent more than 50 % of the reviewed articles.

RQ3: Who is active in this research area?

In order to identify the authors and institutes, who are active in the respective research area, responsible authors and institutes have been linked to each article. The frequency of articles published in relation to the respective authors and institutes can offer a better idea of who is engaged with capability topics in the long-term. Figure 10.5 lists all these countries that published more than three articles the last 15 years.

We identified 30 different countries whereas the USA dominates the list of publication with more than 67 published articles within conferences and journals (2/3 conference articles). Furthermore, it is apparent that German (20) and Chinese (18) research institutes and scientists follow up investigations concerning capability topics. In addition to these very active countries, there are also Australia (17) and Canada (10), which seems to be interested in capability research topics. It is conspicuous that nearly all involved research institutes are resident in the world's leading industrial nations.

Furthermore, the publication activities can be differentiated regarding the publishing institutions. For instance, a drill down of the USA basket of published articles shows an evident tendency that especially in the South-East lots of institutions are dealing with capability topics. Figure 10.6 pictures an extract of institutions that has had more than one article. Most of them are universities like University of Texas, Georgia State University or Emory University. Nevertheless, University of Münster (Germany), City University of Hong Kong, RMIT

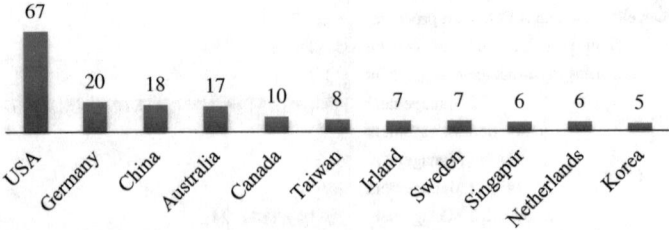

Fig. 10.5 Poblication per country >3

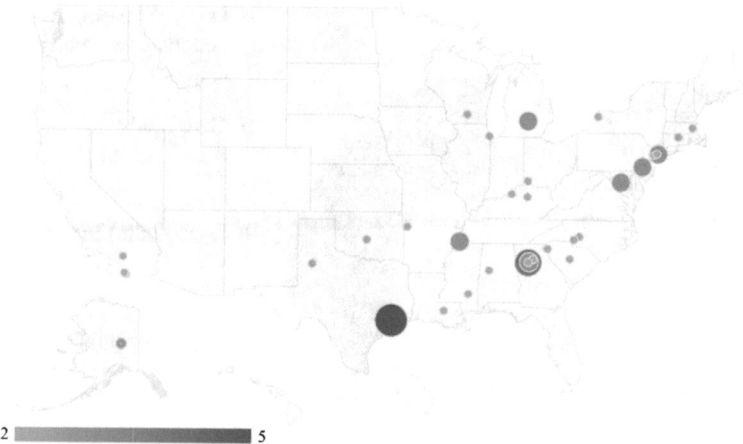

Fig. 10.6 Publication per district in the USA

University (Australia) or Queen's University (Canada) represent additional examples that published at least two articles.

Only a small number of authors belong to a company instead of a university. For instance, we discover companies like Sogeti Netherlands (Dedicated to Technology and Testing Services), Centric (offers software solutions, IT outsourcing, business process outsourcing and staffing services), Z-Sharp (offers IT and business service) or alfabet AG (software company providing an EAM tool + consulting services).

Another interesting fact could be found when looking at the cited authors in articles. Bhardwaj [7], Sambamurthy [31] and Helfat and Peteraf [12], each of them, has been cited more than thousand times. It can be assumed that their work represents recognized scholars in the field. Nevertheless, within this work RQ3 was answered focusing on publishing institutes and corresponding countries. Topics like co-authorship or citation analysis are considered within the review, but not part of this paper.

RQ4. What research approaches are being used?

Beside the article count of involved authors and their institutes, another interesting point of a comprehensive analysis is important. Authors of relevant articles used different research approaches and methods to acquire preferred results. Focused on IS research, Wilde and Hess [38] classified two generally research approaches, firstly, the design science research and secondly, behavioristic research methods. Prototyping, simulation, reference modeling, conceptual -deductive, argumentative deductive analysis and action research are mainly used in design science research. Grounded theory, quantitative-empirical analysis, qualitative-empirical analysis and case studies are behavioristic methods.

Figure 10.7 illustrates that quantitative- and qualitative-empirical analyses are the most common research methods. They are often supported by methods like

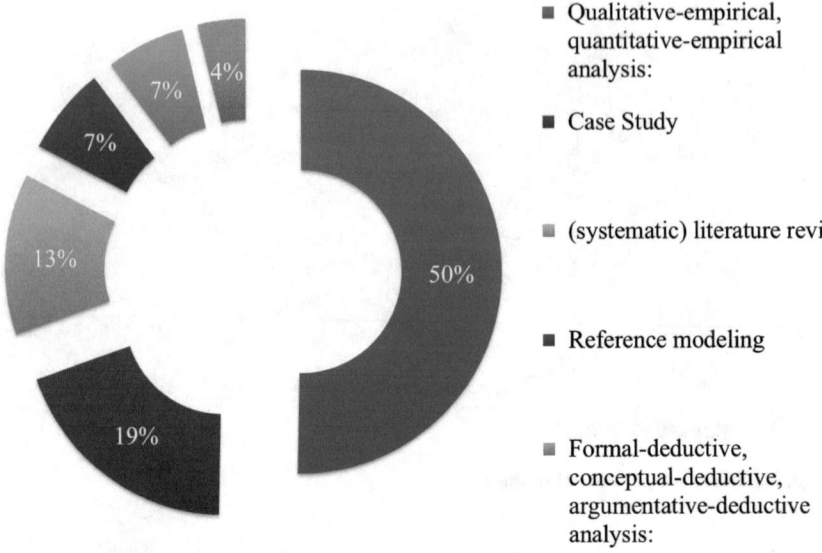

Fig. 10.7 Research approaches

literature analysis or reference modeling. The distribution of research methods illustrates that capabilities have practical relevance, because ~77 % of all research methods were used with practical focus.

RQ5: What kind of capability definitions and descriptive elements are being used?

With RQ2 it has already been stated that relevant articles discussed different research subjects. This situation and the diverse capability interpretations lead to a variety of distinct views. Due to the huge amount of different viewpoints regarding the term "capability", it was obvious to establish a structure based on found definitions and its elements. Organizational capability, (strategic) technological capability, (strategic) business capability, IT knowledge integration, customer orientation capability represent just a small set of identified viewpoints. In order to differentiate capability-types we superimposed all found definitions and looked for characteristics or elements that specify its type more detailed. Similarities have been extracted, sorted and combined to the following capability types (A) and its basic structure in terms of descriptive elements (B).

10.5.1 Types of Capabilities

The range of investigated research subjects contains IT-Management, Business Process Management, Project Management, Supply Chain Management, Knowledge Management and more. We started with a pre-categorization into

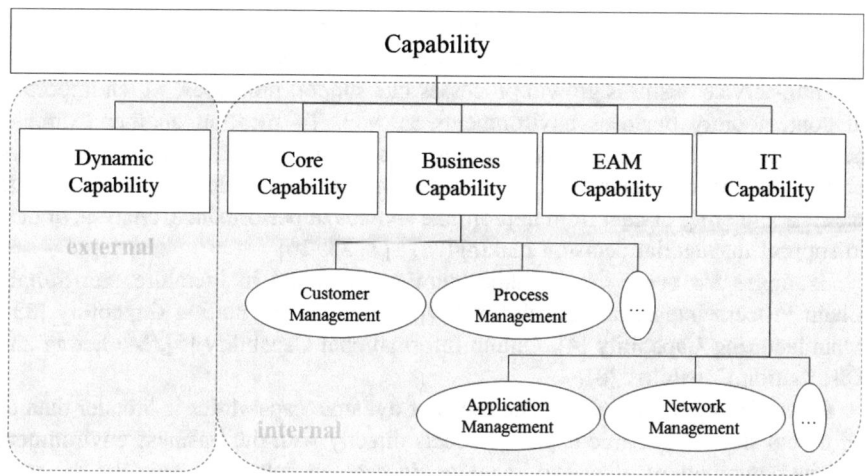

Fig. 10.8 Capability types and possible subsets

internal and external capabilities that is based on Wade and Hulland [37] and its interpretation of how capabilities can be classified. Wade and Hulland speak of inside-out, outside-in, and spanning capabilities while the inside-out capabilities deal with internal affairs, outside-in with external affairs and spanning capabilities involve both internal and external.

Referred to all analyzed articles we found four basic capability cluster (Fig. 10.8). On the one hand the "Business Capability", "IT-Capability" and "Core Capability" that focus on internal operations, and on the other hand "Dynamic Capabilities" that are mainly used within the enterprise's environment. In the following, all capability types are explained in more detail.

IT-Capabilities (117): Using their own IT-capabilities enterprises are able to mobilize IT-resources, "to leverage their IT infrastructure to provide accurate, timely, and reliable data and information to users" [26], and to manage their IT resources in order to realize agility. The central goal of IT-capability represents the realization of business value and maintenance of competitive advantages in terms of IT services and/or IT products. Furthermore, IT-capabilities are used to develop, mediate and leverage other organizational capabilities—e.g., business and core capabilities—and thus, are sometimes described as subtype or subcategory in the literature [29]. The IT Knowledge Integration Capability represents a subtype that concatenates knowledge management and IT resources [6]. Examples for synonyms or subcategories we found in literature are: IS-capability [13], IT infrastructure integration capability [29], IT infrastructure capability [23] or IT Knowledge Integration Capability [6].

Business Capabilities (73): Referred to a corporate business goal the aim of business capabilities is to activate, use and maintain resources for specific business activities. These capabilities may belong to different business management sections as seen in Fig. 10.8. For instance, customer management capabilities enable the detection and determination of requirements and preferences to a company's customer. Process management capabilities are set in product delivery, non-product and non-service business growth processes like support processes, which important in contemporary business environments as well. To mention another example, performance management capabilities are used to "design and manage an effective performance measurement and analysis system, including selection of appropriate metrics, gathering of data from appropriate sources of performance, analysis of data to support managerial decision making[…]" [7, 23, 26].

Examples for synonyms or subcategories we found in literature are: Supply Chain Process Integration Capability [29], Customer Orientation Capability [33], Manufacturing Capability [4], Online Informational Capability [5], Marketing and Distribution Capability [8].

Dynamic Capabilities (73): The focus of dynamic capabilities is broader than of all others since a dynamic capability deals directly with the business environment and its contemporary dynamic behavior. In case an enterprise acquired dynamic capabilities, it has the ability to be responsive to alterations of enterprise environment by e.g., recombining resources. Thus, enterprises are able to identify changes within the environment and to respond to it. Dynamic capabilities are steadily used in combination with other capabilities in order to maximize performance or goals [19]. The Innovative Capability subtype refers to the development and supply of both new products and services.

Core Capabilities (13): Core capabilities are described in general terms. They represent the execution of core competencies within a business process for the purpose of providing either products or services. In addition, core capabilities are supported by both enabling (these capabilities that are necessary but not sufficient) and supplemental (even though they create an added value, they are replaceable) capabilities. Examples for synonyms or subcategories we found in literature are: core BPM capabilities [24].

EAM Capabilities (19): An EAM capability describes the specific combination of know-how in terms of organizational knowledge, procedures and resources able to externalize this knowledge in a specific process with appropriated and available resources to achieve a specific outcome for a defined strategic initiative that change an EA.

10.5.2 Descriptive Capability Elements

Due to the wide spectrum of research subjects and capability-type assignments presented above, it was hard to find a general "capability" explanation in a way that comprises all research subjects and types occurred. At a first glance, we just summarized found capability types and extracted its capability definitions in order to evolve a comprising understanding for it which results in the following definition: "In general a capability involves the ability of an organization to use and combine available tangible and intangible resources to accomplish or enhance business processes and tasks in order to reach predefined goals". Questions like "How (can capabilities be enabled)?", "What (can be done with these capabilities)?" and "Why (is the usage of these capabilities useful)?" can be generally described by using this general description. Nevertheless, with this kind of capability picture we have not made any progress in order to deliver detailed descriptions for different capability types and answer questions like: "What does my organization need to be equipped with an EAM capability like Impact Analysis IS Architecture? What are the key elements of my business capability Customer Management? With this a definition we are not able to answer such questions. Thus, on a second attempt we analyzed relevant articles for potential descriptive capability elements. For example, we start this investigation by analyzing the usage of most obvious descriptive elements of a capability. Amit and Schoemaker already described capabilities as abilities that "[...] refer to an organization's ability to assemble, integrate, and deploy valued resources, usually, in combination or compresences" [2]. They figured out that capabilities are formed and build up on resources, which need to be used in order to do something. Nevertheless, they did not describe the cause why capabilities should be used and what kind of additional aspects like information or activities should be considered. Nineteen years later in another example, Ortbach et al. [28] describe that a capability refer to the ability of an enterprise to perform coordinated activities/tasks (which needs governance) to reach defined goals, which resembles with the definition of a process—maybe the next descriptive element of a capability. Furthermore, they assigned capabilities to resources and assets as well. We listed, counted and aggregated potential descriptive elements from the whole set of relevant articles which results in the following outcomes:

- *Resource* (103): over hundred times a capability was related to tangible/material or non-tangible/immaterial goods that are required in order to define capabilities.
- *Enterprise Context* (84): Capabilities are connected to an overarching subject (see RQ2) or an environment (internal/external) that consider any relevant information which describe the specific situation of an enterprise.
- *Outcome* (101): As an enterprise represents a goal-oriented system, every capability is attended to a certain business goal from a logical perspective. Nevertheless, identified capability concepts are not always directly related to a business goal. In this case, business goals mostly referred to firm performance and competitiveness arguments in terms of outcomes (e.g., produce competitive advantages, satisfying customer wishes, provide services).
- *Processes* (73): 73 times capability was associated to business processes that represent the sequence of activities in order to achieve a certain outcome.
- *Information/Knowledge* (72): 72 times a capability was linked to an information concept that represents a requirement for owning this specific capability. If we identified information and its demand in a specific (enterprise) context, we classified a knowledge demand.
- *Role/Actor* (50): 50 times capability was assigned to some roles or actors. In this case, these roles or actors could be organizational units like marketing, financial and accounting, etc. (i.e. specific domains within an organization).

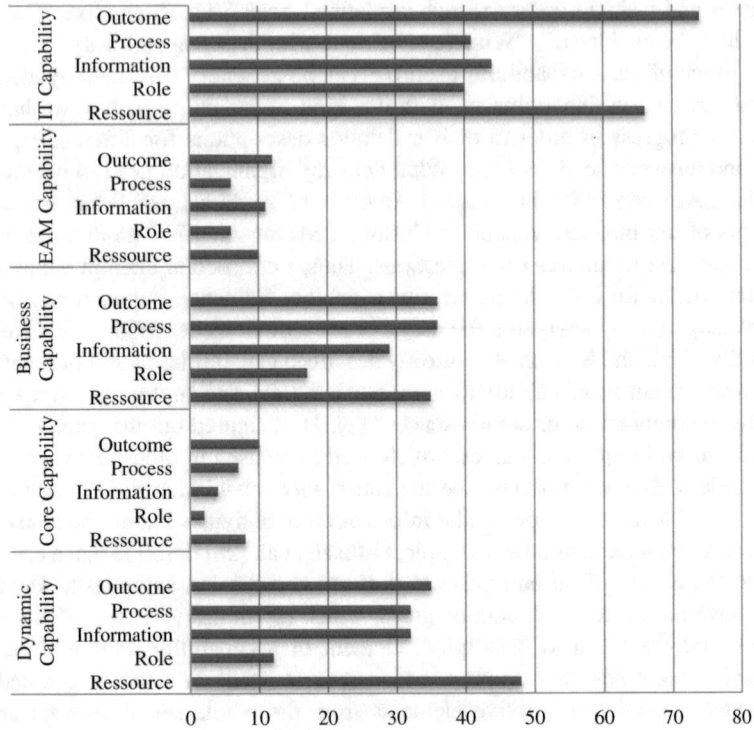

Fig. 10.9 Distribution of descriptive elements by capability type

Figure 10.9 illustrates the distribution of descriptive elements by capability type. For instance, goal and resource oriented element descriptions have often been used to describe the character of an IT capability. The relative distribution of descriptive elements within an IT capability is shown in Fig. 10.10.

All in all the most important element of a capability is represented by a goal definition (64 %), followed by specification of required resources (62 %). Process and Information are associated with 46 % and 45 % to a capability. Last but not least roles are just considered by 29 % (Fig. 10.11).

10.5.3 Correlations of Capability Elements

The analysis of the descriptive elements leads us to a conceptual illustrations based on the correlations (Table 10.2) of capability elements. A capability represents the ability of an enterprise to join information and roles able to execute a specific activity

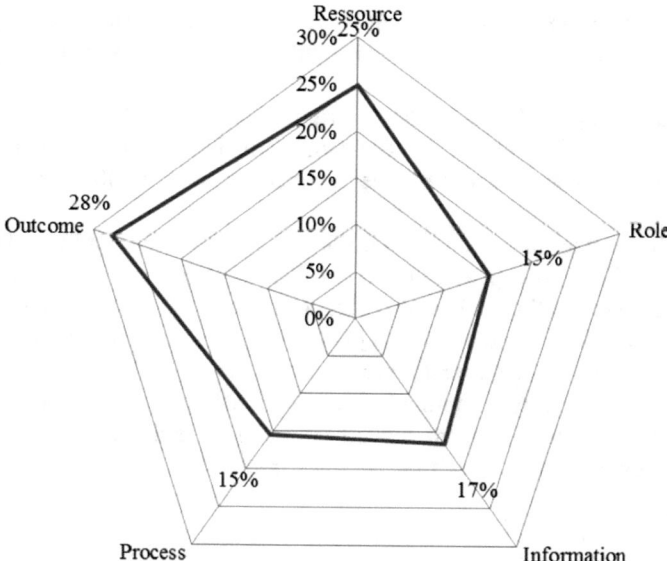

Fig. 10.10 IT capability—distribution of its descriptive elements

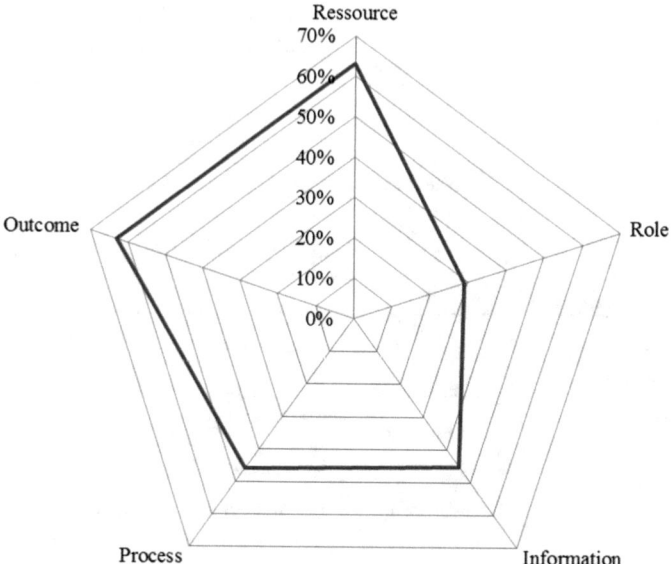

Fig. 10.11 Overall distribution of capability's descriptive elements

with available resources in order to support strategy goals under consideration of its context. We illustrate the relationships and elements of our findings in the Fig. 10.12.

The gray rectangle illustrates the descriptive elements and its interrelations that bring a capability into existence. The number in brackets describes the relative relationship between two descriptive elements based on 508 found relations (Table 10.2).

A capability takes place in a specific context like business-, EAM- or IT-context. The specification of a capability context enhanced its accuracy considering capability management activities like identification, development or maintenance which positively affected its outcome as well [42]. For a specific capability a defined set of roles act on a process (e.g., plan, execute or control). In order to preform its tasks each role is occupied by an optimal set of resources (e.g., competencies or skills). Furthermore, resources (human resources, material, and immaterial goods) are consumed by a set of activities/processes performed by roles as well. That

Table 10.2 Correlations of capability elements

	Resource	Role	Info.	Process	Goal
Resource	–	53	58	58	99
Role	0.10	–	28	30	45
Information	0.10	0.05	–	40	67
Process	0.11	0.06	0.07	–	62
Outcome	0.18	0.08	0.12	0.11	–

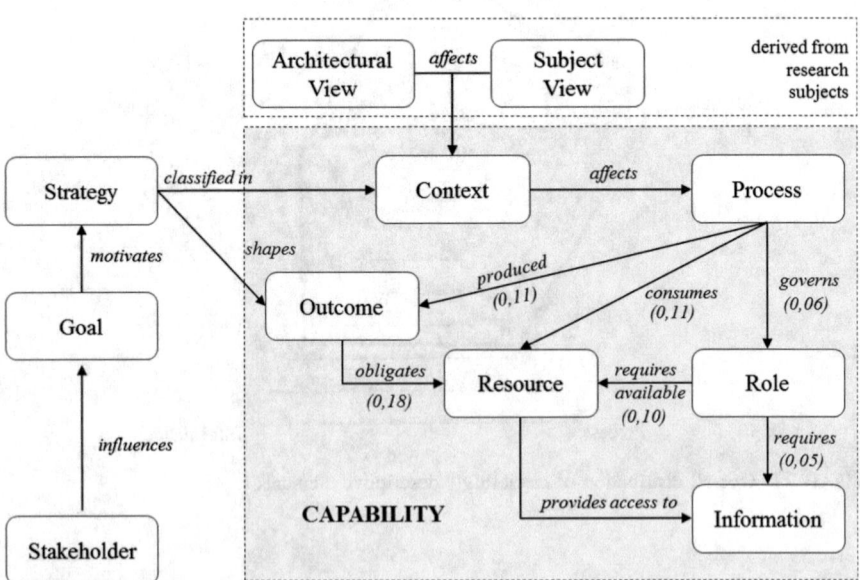

Fig. 10.12 Correlations of capability elements

activities/processes generate the capability desired outcome and could be iterative or divided in sub-processes. Information required for process execution, corresponding roles could be blended of explicit, embodied/implicit or embedded information. The desired outcome of a capability enables the achievement or decisions about the implementation of strategic goals.

10.6 Conclusion and Outlook

Enterprises reach their goals by implementing strategies. Therefore, organizations have to take appropriate actions, which are being summarized by these strategies. A successful strategy implementation is also accompanied by challenges that an enterprise has to face and to overcome. Enterprises require specific capabilities in order to be able to implement strategies efficiently and achieve a specific outcome. We already realized that the capabilities related topics are widely treated in lots of different research areas and publications which motivated us to get a comprehensive overview of the status quo in literature. In order to do so, this investigation merged six systematic literature reviews following the same structural pattern [20]. Therefore we scan five scientific databases and over 21 people analyzed more than 190 relevant articles. We could confirm our impression from recent investigation [42] that an increasing awareness of capability related topics occurs within the last 15 years. A lot of journal articles have been written in the last 5 years and been published in the USA.

The analyzed articles are citied over thirteen thousand times, whereas more than 93 % of all citations are accomplished by 12 different journals representing just a quarter of all relevant articles. It is remarkable to note that the MISQ provides 63 % of all citations, followed by the SMJ with just 22 % (Fig. 10.13). Therefore, the most cited article is published by the MISQ journal as well. "A resource-based perspective on information technology capability and firm performance: an empirical investigation" [7] leads the ranking with more than 3000 citations (captured 08.04.15)

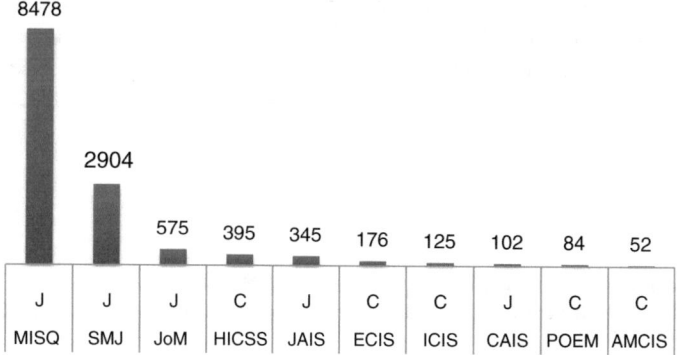

Fig. 10.13 Top ten impact relevant capability sources by citations >50

followed by SMJ article "The dynamic resource-based view: capability lifecycles" [12] with 2229 citations. On the conference side the articles "Development and Validation of a Knowledge Management Capability Assessment Model" [21] and "Developing eInteractions—A Framework for Business Capabilities and Exchanges" [11] are representing the top two cited articles with 62 and 50 citations which already represents 12 % of all conference paper citations (930).

The analysis of research approaches identifies that capability research is bounded on behavioristic research and its methods. The trend of the last few years shows that quantitative-empirical analyses, literature reviews and case studies are suitable methods for capability research. Over the last five years we could identify a trend of multi-methodological research approaches. The usage of different research approaches and methods in a single article seems to deliver more accurate research results and practice-oriented problem solving, which support the utilization of such research approaches like design science as well. However, a number of articles could not be mapped to a research method or method, because it was not described or not comprehensible argued in the text.

The results of this literature exploration end in a conceptualization of a capability term, which includes a set of descriptive elements like outcome, information, role, resource, and activity/process. We added the element context, which represents an additional perspective like application area or subject that we derived from identified research subjects. We identified five capability types, but these are not free from overlaps. The classification of these capability types and its elements should form the basis a capability management approach that supports capability identification, structuring and maintenance in order to enhance strategy implementation quality.

Future work needs to validate these findings in order to define a clear categorization of capability types. Nevertheless, this work is limited by quantitative and qualitative factors. The range of selected literature sources and time frame could be extended in order to discover additional literature sources. Furthermore, fee-based literature (e.g., Gartner Inc., Forrester Inc.) or additional library literature infrastructures could be analyzed in order to expand the set of recognized articles. From a qualitative point of view, the article analysis and classification was performed by a given process (e.g., SLA process, search string, threefold control) and given explanations for used concepts, but individuals perform and interpret information slightly different anyway. Most interpretations above just provide information about the quantity of article contributions and participation in the scientific discourse. We cannot provide a statement or measurement of quality of each contribution.

References

1. AISeL-database. http://aisel.aisnet.org/. Accessed 24 March 2015
2. Amit, R., Schoemaker, P.J.H.: Strategic assets and organizational rent. Strateg. Manag. J. **14** (1), 33–46 (1993)

3. Banker, R.D., Bardhan, I.R., Chang, H., Lin, S.: Plant information systems, manufacturing capabilities, and plant performance. MIS Q. 315–337 (2006)
4. Barua, A., Konana, P., Whinston, A.B., Yin, F.: An empirical investigation of net-enabled business value. MIS Q. **28**(4), 585–620 (2004)
5. Basaglia, S., Caporarello, L., Magni, M.: The mediating role of IT knowledge integration capability in the relationship between team performance and team climate, Italy (2009)
6. Bharadwaj, A.S. (2000) A resource-based perspective on information technology capability and firm performance: an empirical investigation. MIS Q. 169–196
7. Ceccagnoli, M., Forman, C., Huang, P., Wu, D.J.: Co-creation of value in a platform ecosystem: the case of enterprise software. MIS Q. **36**(1), 263–290 (2012)
8. CORE Journal Ranking. http://www.core.edu.au/ (2010). Accessed 24 March 2015
9. CORE Conference Ranking. http://103.1.187.206/core/ (2014). Accessed 24 March 2015
10. EBSCOhost Online Research Databases. http://www.ebscohost.com/. Accessed 24 March 2015
11. Goldkuhl, G., Lind, M.: Developing eInteractions—a framework for business capabilities and exchanges. In: ECIS 2004 Proceedings, vol. 72 (2004)
12. Helfat, C.E., Peteraf, M.A.: The dynamic resource-based view: capability lifecycles. Strateg. Manag. J. **24**(10), 997–1010 (2003)
13. Hobbs, G., Scheepers, R.: Agility in information systems: enabling capabilities for the IT function. Pac. Asia J. Assoc. Inf. Syst. **4**, 2 (2010)
14. Hwang, Y., Kettinger, W.J., Yi, M.: Understanding information behavior and the relationship to job performance. Commun. Assoc. Inf. Syst. **8**, 113–128 (2010)
15. IEEE Xplore Digital Library. http://ieeexplore.ieee.org/Xplore/home.jsp/. Accessed 24 March 2015
16. Journals—Wiley Online Library. http://onlinelibrary.wiley.com/browse/publications?type= journal. Accessed 24 March 2015
17. Journal of Management. http://jom.sagepub.com/. Accessed 24 March 2015
18. Journal Quality List. http://www.harzing.com (2015). Accessed 24 March 2015
19. Kim, G., et al.: IT capabilities, process-oriented dynamic capabilities, and firm financial performance. J. Assoc. Inf. Syst. **7**, 487–517 (2011)
20. Kitchenham, B.: Procedures for performing systematic reviews. Keele, UK, Keele University 33 (2004)
21. Kulkarni, U., Freeze, R.: Development and validation of a knowledge management capability assessment model. In: ICIS 2004 Proceedings, 54 (2004)
22. Langenscheits Handwörterbuch Englisch: Heinz Messinger. Langenscheidt KG, Berlin, Germany (2009)
23. Lu, Y., Ramamurthy, K.: Understanding the link between information technology capability and organizational agility: an empirical examination. MIS Q. **35**, 931–954 (2011)
24. Mathiesen, P., et al.: A critical analysis of business process management education and alignment with industry demand: an Australian perspective. Commun. Assoc. Inf. Syst. **1**, 27 (2013)
25. McLean, E.R., DeLone, W.H.: The delone and mclean model of information systems success: a ten-year update. J. Manag. Inf. Syst. 9–30 (2003)
26. Mithas, S., Ramasubbu, N., Sambamurthy, V.: How information management capability influences firm performance. MIS Q. **35**, 237–256 (2011)
27. Oxford Dictionary Thesaurus & Wordpower Guide. Oxford Univerity Press, New York, USA (2001)
28. Ortbach, K., Plattfaut, R., Poppelbuss, J., Niehaves, B.: A dynamic capability-based framework for business process management: theorizing and empirical application. In: System Science (HICSS), 2012 45th Hawaii International Conference, 4287, 4296 (2012)
29. Rai, A., Patnayakuni, R., Seth, N.: Firm performance impacts of digitally enabled supply chain integration capabilities. MIS Q. 225–246 (2006)
30. Riempp, G., Urbach, N., Smolnik, S.: The state of research on information systems success. Bus. Inf. Syst. **4** (2009)

31. Sambamurthy, V., Bharadwaj, A., Grover, V.: Shaping agility through digital options: reconceptualizing the role of information technology in contemporary firms. MIS Q. 237–263 (2003)
32. SCImago Journal & Country Rank. http://www.scimagojr.com. Accessed 24 March 2015
33. Setia, P., Venkatesh, V., Joglekar, S.: Leveraging digital technologies: how information quality leads to localized capabilities and customer service performance. MIS Q. **37**, 565–590 (2013)
34. Simon, D., Fischbach, K., Schoder, D.: An exploration of enterprise architecture research. Commun. Assoc. Inf. Syst. **32** (2013)
35. Tallman, S., Fladmoe-Lindquist, K.: Internationalization, globalization, and capability-based strategy. Calif. Manag. Rev. **45**, 116 (2002)
36. Tiwana, A., McLean, E.R.: Towards a theory of architectural knowledge integration capability: a test of an empirical model in eBusiness project teams. In: ECIS 2001 Proceedings, vol. 113 (2001)
37. Wade, M., Hulland, J.: Review: the resource-based view and information systems research: review, extension, and suggestions for future research. MIS Q. **28**, 107–142 (2004)
38. Wilde, T. Hess, T.: Forschungsmethoden der Wirtschaftsinformatik. Wirtschaftsinformatik, 280–287 (2007)
39. William U., Rosen, M.: The business capability map: the "Rosetta stone" of business/IT alignment. Enterp. Archit., Cutter Consortium **14**(2) (2011)
40. Wißotzki, M., Koç, H., Weichert, T., Sandkuhl, K.: Development of an enterprise architecture management capability catalog. In: Kobyliński, A., Sobczak, A. (eds.) Perspectives in Business Informatics Research, vol. 158, Springer (2013)
41. Wißotzki, M., Sandkuhl, K.: Elements and characteristics of enterprise architecture capabilities. Perspectives in Business Informatics Research. Springer International Publishing, 82–96 (2015)
42. Wißotzki, M.: The capability management process—finding your way into capability engineering. In: Simon, D., Schmidt, C. (eds.) Business Architecture Management— Architecting the Business for Consistency and Alignment; To be published by Springer in the series "Management Professionals" (2015)
43. SpringerLink. http://link.springer.com/. Accessed 24 March 15

Chapter 11
Enterprise Architecture Analytics and Decision Support

Rainer Schmidt and Michael Möhring

Abstract The discipline of Enterprise Architecture Management started using a model-driven approach. In contrary to the model-driven approaches, our approach follows strives to tap also the information contained in the operational systems that support IT-Service-Management. Therefore, this paper aims at indicating the increased capabilities of Enterprise Architecture Analytics and Decision Support through the use of a data-driven approach. It will give fundamental insights in the current research work of enterprise architecture management analytics as well as decision support based on this quantitative data.

Keywords Enterprise architecture management · IT-Service-Management · Decision support · Analytics

11.1 Introduction

Enterprise architecture management (EAM) has several benefits for enterprises and organizations and is a very important keystone and challenge for modern enterprises [1–4]. EAM is according to Aier et al. [5] "[...] concerned with the establishment and continuous development of EA". In which Enterprise architecture (EA) can be understood as the "the fundamental structures of a company (or government agency) and enables its transformation by bridging the gap between business and information technology (IT)" [5].

The discipline of Enterprise Architecture Management started using a model-driven [6] approach. Based on the enterprise strategy, "to-be" architectures were designed and modeled using notations such as Archimate [7] following a top-down approach. The top-level models were then refined into information

R. Schmidt (✉) · M. Möhring
Munich University of Applied Sciences, Lothstrasse 64, 80335 Munich, Germany
e-mail: Rainer.Schmidt@hm.edu

© Springer International Publishing Switzerland 2016   201
E. El-Sheikh et al. (eds.), *Emerging Trends in the Evolution of Service-Oriented and Enterprise Architectures*, Intelligent Systems Reference Library 111,
DOI 10.1007/978-3-319-40564-3_11

system architectures and application design. However, at the same time there are large collections of data from the operational parts of IT-infrastructure in databases such as the configuration management database [8]. Unfortunately, the models of Enterprise Architecture created in a top-down manner and the data from IT-infrastructure collected in a bottom-up manner are not integrated. This rupture was already identified in [9] and the vision of real-time enterprise architecture was sketched. Also in [10, 11] the need for an integration of enterprise architecture and data from operational IT-Systems has been identified and the vision of an auto-mated integration has been outlined.

The lack of integration between enterprise architecture and operational IT-systems becomes even more urging by the evolution of the IT-infrastructure. Technical architectures vary and applications evolve independently from strategy. As a result, abstract models and IT-Infrastructure develop apart. This phenomenon is known as model-reality-gap [12]. To cope with it, a number of approaches have been developed. In [13] process models are introduced for maintaining enterprise architecture models. They are based on information from both human and technical interfaces. However, the processes are very abstract and first ideas are provided how to operationalize them. In [14] the process designs for automation are joined with others to provide a complete framework of automatic maintenance processes. Other approaches start from the process levels such as [15]. The use of a data warehouse [16] for combining enterprise architecture and operational data is depicted in [17]. The uses of an Enterprise Service Bus for automating enterprise architecture doc-umentation is proposed in [18]. System decision making can be supported by enterprise architecture models and analyses [19].

In contrary to top-down, model-driven approaches, our approach follows strives to tap also the information contained in the operational systems used for IT-Service-Management [20]. Until recently, it was very difficult, expensive, time-consuming or even impossible to collect data for architectural analyzes. Only a small portion could be collected automatically by using specialized software. Most of the data had to be entered manually, a time-consuming and error-prone approach that created huge efforts and costs. The repositories created in this way were often incomplete and erroneous.

Fortunately, the situation has changed due to technological advances. A number of data sources are available as shown in [21]. Today's enterprise architectures such as fabrics [22] and cloud environments [23] are almost entirely composed of vir-tualized resources. This means that the resources used for enterprise architecture are not only customizable but also completely transparent in their structure and their properties. The properties of all computing, storage, and networking elements are queryable and accessible for analysis. Furthermore, many infrastructure compo-nents such as applications, databases etc. are generating information on events, performance resource consumption etc. Modern administration systems such as configuration management databases [8], Log Files [24], and Performance

Monitoring [25] have increased the amount of data available for architectural analysis significantly, too. At the same time, distributed architectures in the context of Big Data [26] such as Hadoop [27] and Spark [28] provide huge-enough computing capabilities to process these data. They also enable the processing of data hitherto ignored or neglected such as semi- and unstructured data.

Therefore, this paper aims at indicating the increased capabilities of Enterprise Architecture Analytics and Decision Support through the use of a data-driven approach. It will give fundamental insights in the current research work of enterprise architecture management analytics as well as decision support based on this quantitative data e.g. [29, 30].

11.2 Data Sources for Enterprise Architecture Analytics

Data for Enterprise Architecture Analytics originates from three classes of sources: structured, semi-structured and unstructured data sources.

11.2.1 Structured Data

Structured data [27] has an explicit schema, that means there are metadata describing the data types and the relationships of the data types. For structured data, a schema-on-write-approach is applied [27]. When data is entered into a database, the schema data is available and the data are organized according to the schema. The most frequent use of structured data are relational databases [31]. They structure data in order to facilitate their use in arbitrary queries. However, there is a tradeoff, the increase in flexibility also increases the effort to read the data. An important example are joins [31] that are used to denormalize data temporarily.

The most important source of structured data for enterprise architecture analytics are fabrics [22]. Fabrics are composed of virtualized computing, storage and networking resources. Using these resources, services are created that support the business process of the enterprise. Fabrics are managed by sophisticated administration systems that manage the whole lifecycle of virtualized resources and the creation of services from them. Examples are the System Center Product Suites from Microsoft [32] or Vmware [33]. The core of these systems are configuration management databases that contain representations of all virtualized resources and the services built from them. In contrary to a physical environment all architectural data are complete and in-sync because they are also used for the operational processes in the fabric. Fabrics also provide easy access to events and performance data, because all components created are under the control of the administration system.

11.2.2 Semi-structured Data

Semi-structured data [27] have a schema, however, this schema is not available explicitly. That means, the schema of semi-structured data can be recovered by applying algorithms. An important example of semi-structured data are so-called log-files Log Files [24]. Typically, the entries in log files consist of a line with information such as data, time, type of event etc. Log files do not have an explicit structure; however, their schema can be reconstructed.

Semi-structured data is used with a schema on read approach [27]. They are stored without information on the schema, but the schema is recovered when the data are read from the data store. This approach has both advantages and disadvantages. As long as the data are not changed and their structure accords with the task to be done, data can be accessed faster than data in a relational database. However, as soon as data have to be changed frequently, the frequent effort to reconstruct the schema surmounts that of normalized data.

11.2.3 Unstructured Data

Unstructured data [27] have no recoverable Data schema. Instead, the information has to be extracted by applying statistical approaches etc. Unstructured data originate from text files, e.g documentation. They contain information that does not follow a predefined schema. Instead, they contain information belonging to different contexts or even completely irrelevant information.

Examples for unstructured data are documents containing descriptions of information systems architectures. They contain the information how the components of the architecture relate to each other. However, this is done in natural language and does not follow a predefined or at least standardized structure. In consequence, every single chunk of information has to be extracted by linguistic analysis.

11.3 Analyzing Architectural Data

Classical data warehouses have been used for the analysis of architecture data since long [16]. However, they are not suitable for the processing of semi—and unstructured data [29]. Although the extract transform load cycle (ETL) is at least partially able to integrate semi—and unstructured data, this approach is very computing intensive. Conventional warehouse architectures are overwhelmed by it. Due to these limited processing capabilities, mainly structured data were used for analyzes in enterprise architecture management. Another problem is that the treatment through a series of intermediate data products creates a high latency from data input to the analysis result [34]. This and the limited use of semi—and unstructured data restrict the use of data warehouses to strategic decision-making. Furthermore, the data analyzed are mostly static ones, describing architectural relations of single items. The analyzes performed were mainly a posteriori ones that created aggregations such as the number of hardware items or software licenses used (Fig. 11.1).

One reason for the hesitant use of semi-structured data before the advent of Big Data [26] has been the computation effort needed to rebuild the schema. Unstructured data require an even higher effort [27] and, therefore, were ignored mostly. It does not surprise that many data from event logs were thrown away after a short time without applying any analysis on them. By the progress of Big Data, the situation has changed significantly. Big Data is a very disruptive information technological development [35, 36]. The advance provide by Big Data can be described in the three dimensions volume, variety and velocity [37] as shown in Fig. 11.2.

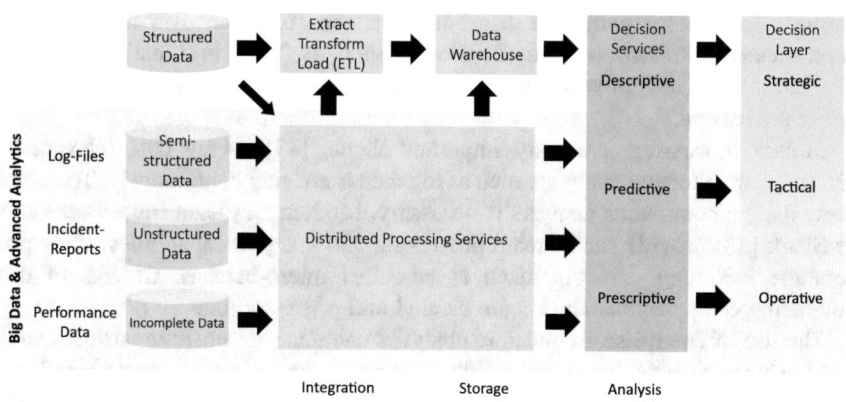

Fig. 11.1 Processing of architectural data based on [29]

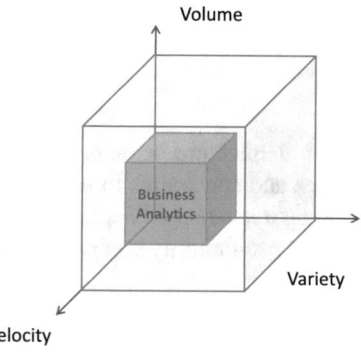

Fig. 11.2 Big data scales out analytics based on [38]

Big Data significantly increases the type and amount of data available for enterprise architecture analytics [29]. Now not only backward-looking, descriptive analyzes can be done, but also forward-directed predictive and prescriptive analyzes. Therefore, not only strategical decisions but also tactical and operational decisions are supported.

The capabilities of Big Data process and increased volume, variety and velocity of data is based on its highly distributed processing architecture [27]. It enables increasing the volume of data by processing data in parallel. Also, the latency of computation is decreased, that means the velocity of computation is raised in this way. Furthermore many Big Data approaches use a drive-through approach for processing, avoiding latency-increasing intermediate data such as often found in data warehouses [39]. The increased processing capabilities also enable Big Data to process semi- and unstructured data that need a processing overhead in order to reconstruct the schema, as with semi-structured data, or to detect information in unstructured data.

In the beginning, Big Data frameworks such as Hadoop provided only simple processing algorithms such as MapReduce [40]. It performs transformations and sorting of key-value pairs in a distributed manner. However, over the time more sophisticated frameworks were developed such as Yarn [41] and Spark [28], GraphX [42]. They provide more sophisticated processing capabilities such as graph processing.

Stream processing is a very important theme [43]. Many data relevant for enterprise architecture analytics such as log data is arriving continuously. To exploit these data, a continuous analysis is necessary. Modern Big Data frameworks such as Spark [28] provide such stream processing and analysis capabilities. They process the incoming data organized as so-called micro-batches. Chunks of data embracing of approximately 1 s are created and processed one by one.

The use of enterprise architecture analytics can generate positive business value and business impacts. Important architecture decisions can now be done based on a better data quality and not only on a gut instinct. Therefore the following metrics [44, 45] can be better acquired:

- Cost metrics
- Scalability metrics
- Portability metrics
- Security metrics
- Etc.

A comparison of different architecture variants based on metrics shows CIOs a better view of the possibilities and constraints of each. As a result, better decisions can be applied [29]. This comparison can be applied for example by using a utility analysis [46]. An example of a group utility analysis based on the metrics above is shown in Fig. 11.3.

	Architecture A	Architecture B	Architecture C	Weight	Count
Cost					
Scalability					
Portability					
Security					
...	...	...	...	...	...

Fig. 11.3 Example of utility analysis of architecture variants [29]

11.4 Applications of Big Data and Advanced Analytics in Enterprise Architecture

Big Data and Advanced Analytics offer a number of new application areas in Enterprise Architecture. Three examples shall be presented. The following examples are made in the area of EAM and IT service management [20].

11.4.1 Forecasting the Demand and Prices of EA Services

Schmidt et al. [29] argue, that the IT service demands can be better predicted trough a broader database. Therefore, different metrics and influence factors, as well as information through the analysis of unstructured data (cf. Sect. 11.2), can be used to decrease the forecast error [29]. Knowing the IT demand is very important to construct a stable enterprise architecture. In general the forecast error can be defined as follows [29, 47, 48]:

$$\text{forecast error} = \left| \text{demand}_{\text{real}} - \text{demand}_{\text{predicted}} \right|$$

Equation 1: Forecast error [29]

Furthermore, the quality of the predicted demand can be measured by other metrics like the root mean square error (RMSE), the mean absolute percentage error (MAPE) or the Nash-Sutcliffe coefficient of efficiency (NSC) [49]. Schmidt et al. [29] define a saving of idle time (provided IT services, but not really used) by reducing the forecast error for each IT service demands and pre-production costs.

In general, there are different quantitative approaches to predict time series [50, 51] like autoregressive integrated moving average models (ARIMA), linear regression, artificial neuronal networks or Winters/Holt methods. The basic element of artificial networks is a neuron [52, 53]. This has a number of inputs. Based on weighted inputs, the neuron computes its output. Several neurons can be organized into layers, which can be switched again in a row. There are different types of neuronal networks. Some are connected in one direction only and others allow a feedback from the output to the input. The peculiarity of deep learning lies in the

Sub-symbolic way of working [54]. While classic machine learning techniques using symbolic representations of numbers, statements, etc., know that neuronal networks only input and output signals. Particular, artificial neuronal networks are very flexible because they do not ignore the non-linear behavior of the independent variables [51, 55]. Furthermore, trends and structural interruptions must be recognized [56, 57].

For all cases, different prediction algorithms should be tested with historical data and compared by different quality metrics like the forecast error (cf. see above). After evaluation, the best algorithms should be used to predict the IT demand. Prediction algorithms are implemented through many software tools like R project, IBM SPSS, SAS or Rapid Miner.

In the following a short example will be described for prediction the IT demand via Rapid Miner under use of a prediction algorithm. Rapid Miner is a well-known and recognized software for data analytics in practice and research [58]. There are different software license possibilities for Rapid Miner—like a use for free (cf. www.rapidminer.com).

The sample case will be the monthly prediction of the execution of the IT service 4242 ("VM installation"). The IT demand can be used to evaluate the current enterprise architecture and to improve it based e.g. on the service requirements. The amount of requests and the firm's turnover is given for execution (Table 11.1).

Different prediction algorithms can be used in Rapid Miner [58]. For instance, an artificial neuronal network for prediction the IT demand can be implemented like shown in Fig. 11.4.

First historical data must be loaded and analyzed by a neuronal network (Neuronal Net). In this step, a model is built for prediction and used by the "Apply Model" component to predict new months with the given independent variables (in our example e.g. requests, turnover, old executions).

In our example, we want to predict the amount of execution of the service 4242 by using historical data of the amount of execution, requests, and turnover one month ago (see table above). By applying an artificial neuronal network algorithm like described above, we got the following results (prediction (Y): predicted amount of execution, X1 last month execution, X2 last month requests, X3: last month turnover) (Fig. 11.5).

Table 11.1 IT demand prediction example

Month	IT-service	IT-service-ID	Executions	Requests	Firms turnover (Mio)
January 2015	VM installation	4242	60	80	4.5
February 2015	VM installation	4242	56	76	5.2
March 2015	VM installation	4242	65	92	5.4
April 2015	VM installation	4242	73	108	5.6
May 2015	VM installation	4242	78	123	5.7
June 2015	VM installation	4242	?	?	?

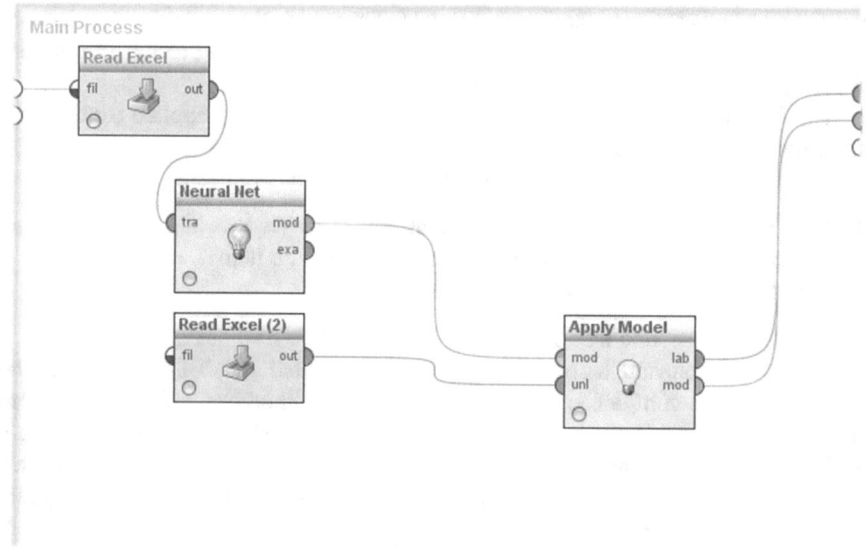

Fig. 11.4 Use of prediction algorithms in Rapid Miner

Row No.	Y	prediction(Y)	X1	X2	X3
1	?	79.204	78	123	5700

Fig. 11.5 Sample prediction via ANN

Other algorithms are easy to implement by change the ANN component to another (e.g. regression, SVM, etc.). Based on prediction evaluation with historical data, the best algorithms should be used.

11.4.2   Service Recommendation for Different Customers

In large enterprises and organizations with a huge number of internal and maybe external IT service customers, it is not easy to offer all customers all of the implemented services. Requirements and use cases maybe differ from department to department as well as from firm to firm. A good IT support and exchange of requirements is very important to well implement IT-Business Alignment [59] to achieve business and IT goals.

Therefore, traditional Data Mining techniques like association algorithms can help to find similarities in the use and buying behavior of IT services in the enterprise architecture of different customers. Association algorithms are very well known for a so-called "market basket analysis", where co-occurrences of items in individual purchases were analyzed [60, 61]. Therefore, the enterprise architecture can be improved by providing a better basement for business process in different departments by analyzing similar user behavior.

The following example based on IT services. It can be also applied for different enterprise architecture patterns as well as architectural decisions.

To generate this information from data several steps are needed [61]:

1. Data preparation in a transaction format
2. Generate binomial format with the occurrence of the items
3. Generation of relevant association rules

The first and second step is illustrated in Fig. 11.6.

Finally the relevant association rules are calculated [60, 61]. Therefore, the Furthermore, the confidence as a measure of the likelihood of support of an item (relative frequency) in the transaction set is extracted [61]. In our short example a VM installation is in all cases (2/2 → support = 1), GSM update in one of two (1/2 → support = 0.5) and the combination of VM installation and GSM update also in one of two cases (1/2 → support = 0.5).

Furthermore, the confidence as a measure of the likelihood of the occurrence is calculated [61]. For our example the following equation describes the calculation of the confidence of getting a GSM update after VMInstallation (confidence = 0.5/1 → 0.5) [61]:

$$
\begin{aligned}
Confidence\ & VMInstallation \rightarrow GSMupdate \\
& = \frac{Support(VMInstallation \cup GSMupdate)}{Support(VMInstallation)}
\end{aligned}
$$

Equation 2: sample of the calculation of confidence (according to [61])

IT service order number	Ordered services
815	VM Installation, GSM update
816	VM Installation
...	...

to binomial

IT service order number	VM Installation	GSM update	...
815	1	1	
816	1	0	
...	...	...	

Fig. 11.6 Item set generation

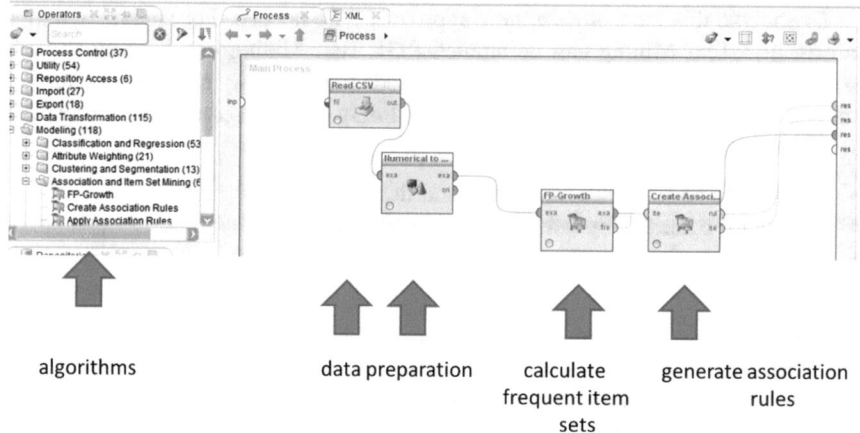

Fig. 11.7 Rapid Miner for association rules generation

Finally the relevant association rules are calculated [60, 61]. Therefore, the support of an item (relative frequency) in the transaction set is extracted [61]. There are further metrics like lift or conviction to analyze association rules as well as different processes of rule generation implemented in Text Mining software [61]. For instance in Rapid Miner [58] it is possible to generate association rules with the "association and item set mining" part (Fig. 11.7).

First, data is loaded (e.g. via CSV). After this step, binomial item sets are calculated. The third step is the calculation of frequent item sets and finally association rules are generated based on this results.

Finally we get the association rules for EA IT services to be more competitive by providing more customer oriented services and adopt as well transform the enterprise architecture to a more customer-oriented one In our example we got the association rule:

Rule 1: VM installation—implies → GSM update

In general, the cases are more complex. This example should only demonstrate the functionality of the algorithms behind this approach. It can also be used for EA patterns (in contrast to IT services).

11.4.3 Analyzes of Unstructured Data

In the field of EAM, there are a lot of unstructured data sources [62]. For instance, documentation in MS Word, PDF or E-Mails as well as descriptions trough EAM software and frameworks.

To analyze this kind of data, new approaches are needed. For analyzing textual information Text Mining can be applied [63]. Text Mining can be defined as an extension of the traditional data mining methods to extracting interesting patterns or knowledge from textual data [63–66]. Text Mining can be used for improving EAM in different ways (cf. [29, 62, 66]):

- check differences between models and documentation for auditing
- improving forecasts by using textual information
- EAM process improvement
- etc.

In general Fig. 11.8 summarizes the possibilities and steps to analyze unstructured data.

After loading textual data (e.g. in PDF or MS Word format) in the Text Mining environment (e.g. R Project or Rapid Miner), the text must be transformed in different tokens (e.g. each word become a token) [67]. Then there is may be a case conversion (e.g. lower case), that each token has the same case [67]. Furthermore (if needed) stop words like "the", "that" can be removed. Finally, the tokens are stemmed to their "basic form". These steps are summarized in Fig. 11.9 [67].

There are different software tools (free or commercial use) to analyze textual data. Rapid Miner is one well-known and established data mining tools [58]. Under use of the Text Mining add-on, textual data for EAM can be preprocessed (Fig. 11.10).

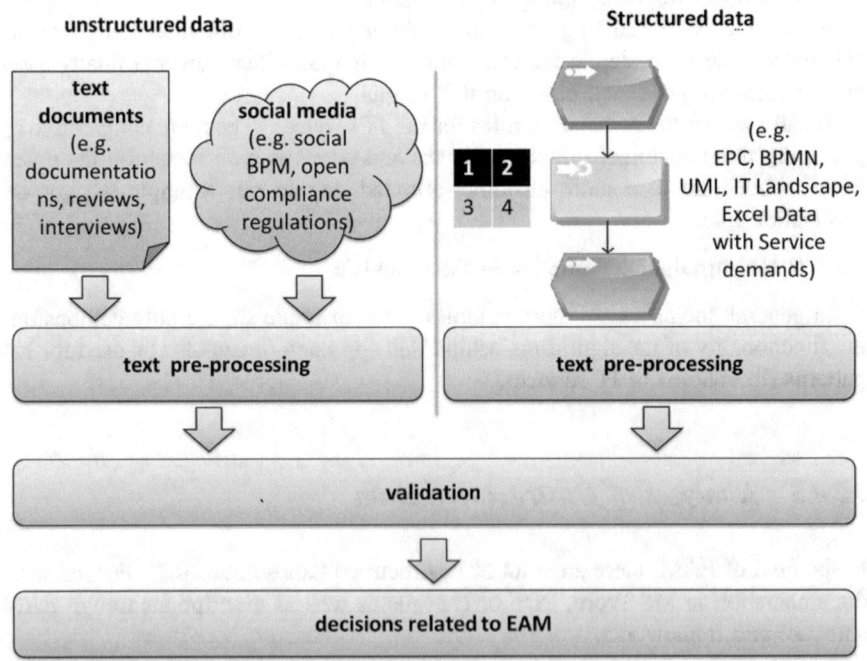

Fig. 11.8 Analyzing unstructured data (according to [66])

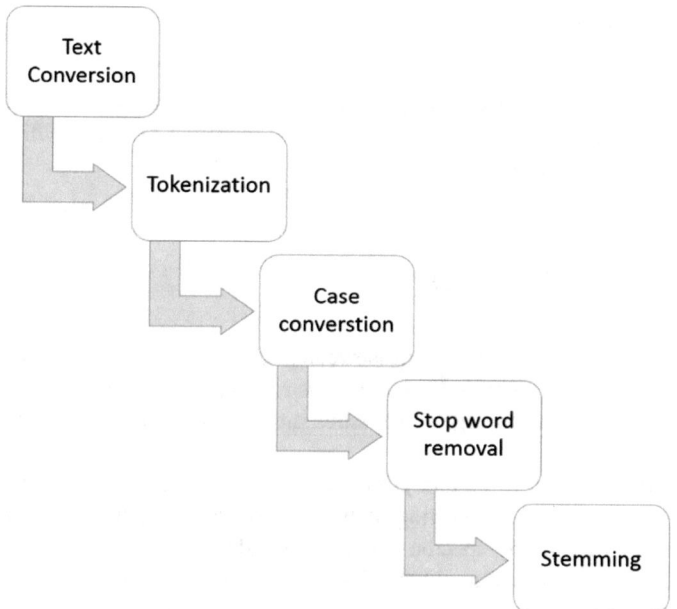

Fig. 11.9 Text pre-processing according to [67]

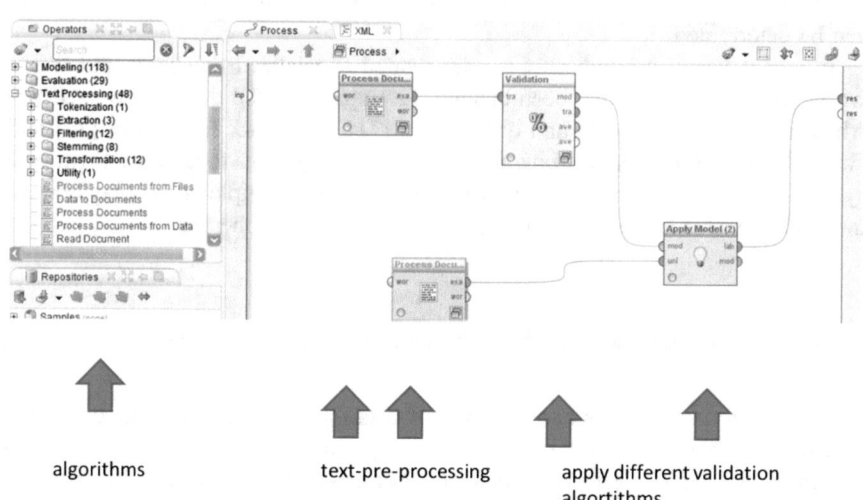

Fig. 11.10 Text Mining via Rapid Miner

First, the text is pre-processed like described above (e.g. tokenization, etc.). Then different classification algorithms (like K-NN, Bayes, SVM) can be used to analyze hidden structure in the text data (e.g. differences between different textual EAM landscape description) [66].

11.5 Outlook

The data-driven approach of enterprise architecture management will profit from a number of technologies in development. Using them the analyzes will be accelerated and improved in quality.

11.5.1 Graph-Based Data

The data in the structure an IT infrastructure can be represented in the form of a directed multi-property graph. Each device is represented as a node. Between nodes may be zero, one or more edges. Properties may be associated with both nodes and edges.

This directed graph to be analyzed using developed and adapted graph mining algorithms to recognize recurring patterns. So, problematic structures can be determined by identifying specific subgraph. In addition, it can be determined how far the current infrastructure differs from risky configurations. To do this, algorithms and methods can be developed, which are based on a modified Graph version of the Levenshtein [68] distance. E.g. by measuring this modified Levenshtein distance between the graph of the present infrastructure and dangerous fabric configurations, important triggers for improving the resilience of the fabric can be determined.

In recent years, a number of graph-oriented specialized databases and frameworks in recent years have been created. Examples are the database Neo4J [69], which can store graph data and query them [71], the Spark framework [70] with GraphX [42] that is capable of parallel processing graph data to a large extent. Both systems can represent multigraphs, i.e. graphs where multiple edges may connect the nodes. Nodes and edges can also be assigned with properties.

11.5.2 Frameworks for Stream-Based Data Processing

Stream-based data [43] are another form of data that often arises in the context of enterprise architecture. Log file entries and performance data are a continuous stream of data, which needs to be analyzed. For this purpose, a series of frameworks is also available. In Spark [70], stream processing is based on micro batch-based processing. The streams of data are divided up into chunks of approximately one-half to several seconds duration. The collected data are processed in turn with the resilient distributed datasets [70]. The preprocessing can be, for example, with Apache Kafka [71] framework. An alternative framework represents the Apache Flink [72]. It offers a full stream processing, i.e. there is no distribution in micro batches like in spark. Contrary to Apache Spark, Flink is not resilient. In the case of malfunctions, lost data are not restored.

11.6 Conclusion

There are two approaches for Enterprise Architecture Management, a model-driven and a data-driven approach. The model-driven approach has long been predominant, because the automatic data collection necessary for the data-driven approach, was very limited. This changed significantly by the increasing proliferation of virtualized environments and big data. Architecture data available are now largely. They can be distinguished according to topology, variability, modality and morphology. Big data and advanced analytics provide the necessary means to analyze this newly available data.

Enterprise architecture analytics leverage a broader database of structured and unstructured data for analyzing the past, current and future enterprise architecture. Therefore, enterprises, as well as organizations, can get a better view of their enterprise architecture and improve decisions related to EAM. This chapter shows the basics of EAM analytics and shows some different practical scenarios how to use EAM analytics.

References

1. Jonkers, H., Lankhorst, M.M., ter Doest, H.W., Arbab, F., Bosma, H., Wieringa, R.J.: Enterprise architecture: management tool and blueprint for the organisation. Inf. Syst. Frontiers. **8**, 63–66 (2006)
2. Schmidt, R., Möhring, M., Härting, R.-C., Reichstein, C., Zimmermann, A., Luceri, S.: Benefits of enterprise architecture management—insights from European experts. Presented at the PoEM 2015: 8th IFIP WG 8.1 working conference on the Practice of Enterprise Modelling, Berlin (2015)
3. Buckl, S., Ernst, A.M., Lankes, J., Matthes, F., Schweda, C.M.: Enterprise architecture management patterns–exemplifying the approach. In: Enterprise Distributed Object Computing Conference, 2008. EDOC'08. 12th International IEEE. pp. 393–402. IEEE (2008)
4. Aier, S., Riege, C., Winter, R.: Unternehmensarchitektur-Literaturüberblick und Stand der Praxis. Wirtschaftsinformatik **50**, 292–304 (2008)
5. Aier, S., Gleichauf, B., Winter, R.: Understanding enterprise architecture management design-an empirical analysis. In: Wirtschaftsinformatik, p. 50 (2011)
6. Power, D.J., Sharda, R., Burstein, F.: Decision support systems. Wiley Online Library (2002)
7. Lankhorst, M.M., Proper, H.A., Jonkers, H.: The architecture of the ArchiMate language. In: Enterprise, Business-Process and Information Systems Modeling, pp. 367–380 (2009)
8. Brenner, M., Garschhammer, M., Sailer, M., Schaaf, T.: CMDB-yet another MIB? On Reusing Management Model Concepts in ITIL Configuration Management. Large Scale Management of Distributed Systems, pp. 269–280 (2006)
9. ter Doest, H., Lankhorst, M.: Tool Support for Enterprise Architecture-A Vision. Telematica Instituut, Enschede (2004)

10. Buckl, S., Matthes, F., Schweda, C.M.: Future Research Topics in Enterprise Architecture Management—A Knowledge Management Perspective. In: Dan, A., Gittler, F., Toumani, F. (eds.) Service-Oriented Computing. ICSOC/ServiceWave 2009 Workshops. pp. 1–11. Springer Berlin Heidelberg (2010)
11. Roth, S., Matthes, F.: Future research topics in enterprise architectures evolution analysis. In: Software Engineering (Workshops), pp. 201–206 (2013)
12. Erol, S., Granitzer, M., Happ, S., Jantunen, S., Jennings, B., Johannesson, P., Koschmider, A., Nurcan, S., Rossi, D., Schmidt, R.: Combining BPM and social software: contradiction or chance? J. Softw. Maintenance Evol. Res. Pract. **22**, 449–476 (2010)
13. Farwick, M., Agreiter, B., Breu, R., Ryll, S., Voges, K., Hanschke, I.: Automation processes for enterprise architecture management. In: Enterprise Distributed Object Computing Conference Workshops (EDOCW), 2011 15th IEEE International, pp. 340–349. IEEE (2011)
14. Farwick, M., Schweda, C.M., Breu, R., Hanschke, I.: A situational method for semi-automated enterprise architecture documentation. Softw. Syst. Model. 1–30 (2014)
15. Correia, A., Abreu, F.: Integrating it service management within the enterprise architecture. In: Fourth International Conference on Software Engineering Advances, 2009. ICSEA'09, pp. 553–558 (2009)
16. Kimball, R., Ross, M., et al.: The Data Warehouse Toolkit: The Complete Guide to Dimensional Modelling. Wiley, New York [ua] (2002) (Nachdr)
17. Veneberg, R.K.M., Iacob, M.E., Van Sinderen, M.J., Bodenstaff, L.: Enterprise architecture intelligence: combining enterprise architecture and operational data. In: Enterprise Distributed Object Computing Conference (EDOC), 2014 IEEE 18th International, pp. 22–31 (2014)
18. Buschle, M., Ekstedt, M., Grunow, S., Hauder, M., Matthes, F., Roth, S.: Automating enterprise architecture documentation using an enterprise service bus (2012)
19. Johnson, P., Ekstedt, M.: Enterprise architecture: models and analyses for information systems decision making (2007)
20. Galup, S.D., Dattero, R., Quan, J.J., Conger, S.: An overview of IT service management. Commun. ACM **52**, 124–127 (2009)
21. Farwick, M., Breu, R., Hauder, M., Roth, S., Matthes, F.: Enterprise architecture documentation: Empirical analysis of information sources for automation. In: 2013 46th Hawaii International Conference on System Sciences (HICSS), pp. 3868–3877. IEEE (2013)
22. Bär, F., Schmidt, R., Möhring, M.: Fabric-Process Patterns. In: Bider, I., Gaaloul, K., Krogstie, J., Nurcan, S., Proper, H.A., Schmidt, R., Soffer, P. (eds.) Enterprise, Business-Process and Information Systems Modeling, pp. 139–153. Springer, Berlin (2014)
23. Schmidt, R.: A framework for comparing cloud-environments. In: 2011 Federated Conference on Computer Science and Information Systems (FedCSIS), pp. 553–556. IEEE, Stettin (2011)
24. List of Log Files in Configuration Manager: (2007) http://technet.microsoft.com/en-us/library/bb892800.aspx
25. Fensterer, M.: Supporting capacity planning of cloud computing data centers with long term trend analysis of performance monitoring data (2012)
26. Zikopoulos, P., Eaton, C.: Understanding Big Data: Analytics for Enterprise Class Hadoop and Streaming Data. McGraw-Hill Osborne Media (2011)
27. White, T.: Hadoop: The definitive guide. O'Reilly Media (2012)
28. Zaharia, M., Chowdhury, M., Franklin, M.J., Shenker, S., Stoica, I.: Spark: cluster computing with working sets. In: Proceedings of the 2nd USENIX conference on Hot topics in cloud computing, pp. 10–10 (2010)
29. Schmidt, R., sotzki, M.W., Jugel, D., Möhring, M., Sandkuhl, K., Zimmermann, A.: Towards a framework for enterprise architecture analytics. In: Grossmann, G., Hallé, S., Karastoyanova, D., Reichert, M., Rinderle-Ma, S. (eds.) 18th IEEE International Enterprise Distributed Object Computing Conference Workshops and Demonstrations, EDOC Workshops 2014, Ulm, Germany, 1–2 Sep 2014, pp. 266–275. IEEE Computer Society (2014)
30. Schmidt, R., Zimmermann, A., Möhring, M., Jugel, D., Bär, F., Schweda, C.M.: Social-software-based support for enterprise architecture management processes. In: Fournier, F., Mendling, J. (eds.) Business Process Management Workshops—BPM 2014

International Workshops, Eindhoven, The Netherlands, 7–8 Sep 2014, Revised Papers, pp. 452–462. Springer (2014)

31. Codd, E.F.: Relational completeness of data base sublanguages. IBM Corporation (1972)
32. Beaumont, S., Gasser, D., Baumgarten, A.: Microsoft System Center 2012 Service Manager Cookbook. Packt Publishing, Birmingham (2012)
33. Bunch, C.: Automating vSphere with VMware vCenter Orchestrator. VMware Press (2012)
34. Hajlaoui, J.E., Hamdani, N.: Active data warehouse: Review, challenges and issues. In: 2014 World Symposium on Computer Applications and Research (WSCAR), pp. 1–6. IEEE (2014)
35. Bughin, J., Chui, M., Manyika, J.: Clouds, big data, and smart assets: Ten tech-enabled business trends to watch. McKinsey Quarterly. 56 (2010)
36. LaValle, S., Lesser, E., Shockley, R., Hopkins, M.S., Kruschwitz, N.: Big data, analytics and the path from insights to value. MIT Sloan Manage. Rev. 52, 21–32 (2011)
37. Chen, H., Chiang, R.H., Storey, V.C.: Business intelligence and analytics: from big data to big impact. MIS Q. 36, 1165–1188 (2012)
38. Schmidt, R., Möhring, M.: Strategic alignment of cloud-based architectures for big data. In: Proceedings of the 17th IEEE International Enterprise Distributed Object Computing Conference Workshops (EDOCW). Vancouver, Canada (2013)
39. Mohanty, S., Jagadeesh, M., Srivatsa, H.: Big Data Imperatives: Enterprise "Big Data" Warehouse, "BI" Implementations and Analytics. Apress (2013)
40. Dean, J., Ghemawat, S.: MapReduce: simplified data processing on large clusters. Commun. ACM 51, 107–113 (2008)
41. Murthy, A.: Apache Hadoop YARN: moving beyond MapReduce and batch processing with Apache Hadoop 2. Pearson, Upper Saddle River, NJ (2014)
42. Xin, R.S., Crankshaw, D., Dave, A., Gonzalez, J.E., Franklin, M.J., Stoica, I.: GraphX: unifying data-parallel and graph-parallel analytics. arXiv:1402.2394 [cs] (2014)
43. Psaltis, G.: Streaming Data. Manning (2015)
44. Aier, S., Ahrens, M., Stutz, M., Bub, U.: Deriving SOA evaluation metrics in an enterprise architecture context. In: Service-Oriented Computing-ICSOC 2007 Workshops, pp. 224–233 (2009)
45. Vasconcelos, A., Sousa, P., Tribolet, J.: Information system architecture metrics: an enterprise engineering evaluation approach. Electron. J. Inf. Syst. Eval. 10, 91–122 (2007)
46. Weirich, P.: Decision space: Multidimensional utility analysis. Cambridge University Press (2001)
47. Leitch, G., Tanner, J.E.: Economic forecast evaluation: profits versus the conventional error measures. Am. Econ. Rev. 580–590 (1991)
48. Cao, L., Soofi, A.S.: Nonlinear deterministic forecasting of daily dollar exchange rates. Int. J. Forecast. 15, 421–430 (1999)
49. Faruk, D.Ö.: A hybrid neural network and ARIMA model for water quality time series prediction. Eng. Appl. Artif. Intell. 23, 586–594 (2010)
50. Vogel, J.: Prognose von zeitreihen. Springer (2014)
51. Zhang, G.P.: Time series forecasting using a hybrid ARIMA and neural network model. Neurocomputing 50, 159–175 (2003)
52. Jain, A.K., Mao, J., Mohiuddin, K.M.: Artificial neural networks: A tutorial. Computer, pp. 31–44 (1996)
53. Zurada, J.M.: Introduction to Artificial Neural Systems. West St, Paul (1992)
54. Schmidhuber, J.: Deep learning in neural networks: An overview. Neural Networks 61, 85–117 (2015)
55. Zhang, G., Patuwo, B.E., Hu, M.Y.: Forecasting with artificial neural networks: The state of the art. Int. J. Forecast. 14, 35–62 (1998)
56. Chow, G.C.: Tests of equality between sets of coefficients in two linear regressions. Econometrica J. Econometric Soc. 591–605 (1960)
57. Hansen, B.E.: Testing for parameter instability in linear models. J. Policy Model. 14, 517–533 (1992)

58. Hofmann, M., Klinkenberg, R.: RapidMiner: Data Mining Use Cases and Business Analytics Applications. CRC Press (2013)
59. Luftman, J., Kempaiah, R.: An update on business-IT alignment: "A line" has been drawn. MIS Q. Executive **6**, 165–177 (2007)
60. Agrawal, R., Imieliński, T., Swami, A.: Mining association rules between sets of items in large databases. In: ACM SIGMOD Record, pp. 207–216. ACM (1993)
61. Kotu, V.: Predictive Analytics and Data Mining: Concepts and Practice with Rapidminer. Elsevier, Waltham (2014)
62. Möhring, M., Schmidt, R., Härting, R.-C., Bär, F., Zimmermann, A.: Classification Framework for Context Data from Business Processes. In: Fournier, F., endling, J. (eds.) Business Process Management Workshops—BPM 2014 International Workshops, Eindhoven, The Netherlands, 7–8 Sep 2014, Revised Papers. pp. 440–445. Springer (2014)
63. Tan, A.: Text Mining: the state of the art and the challenges. In: Proceedings of the PAKDD 1999 Workshop on Knowledge Disocovery from Advanced Databases, pp. 65–70 (1999)
64. Fayyad, U., Piatetsky-Shapiro, G., Smyth, P.: From data mining to knowledge discovery in databases. AI Mag. **17**, 37 (1996)
65. Simoudis, E.: Reality check for data mining. IEEE Intell. Syst. **11**, 26–33 (1996)
66. Schmidt, R., Möhring, M., Härting, R.-C., Zimmermann, A., Heitmann, J., Blum, F.: Leveraging textual information for improving decision-making in the business process lifecycle. In: Neves-Silva, R., Jain, L.C., Howlett, R.J. (eds.) Intelligent Decision Technologies. Sorrent (2015)
67. Tan, P.-N., Blau, H., Harp, S., Goldman, R.: Textual data mining of service center call records. In: Proceedings of the sixth ACM SIGKDD international conference on Knowledge discovery and data mining, pp. 417–423. ACM (2000)
68. Levenshtein, V.I.: Binary codes capable of correcting deletions, insertions, and reversals. In: Soviet physics doklady, pp. 707–710 (1966)
69. Jordan, G.: Practical Neo4j. Apress, Berkeley (2014)
70. Ryza, S. (ed.): Advanced Analytics with Spark: Paterns for Learning from Data at Scale. O'Reilly, Beijing (2015)
71. Kreps, J., Narkhede, N., Rao, J.: Kafka: A distributed messaging system for log processing. In: Proceedings of the NetDB (2011)
72. Markl, V.: Breaking the chains: On declarative data analysis and data independence in the big data era. Proceedings of the VLDB Endowment. **7**, 1730–1733 (2014)

Chapter 12
A Guide for Capability Management

Matthias Wißotzki and Anna Sonnenberger

Abstract Digitalization, shorter product cycles, oversupply of markets and the increasing customer requirements both determine and affect the movement from an industrial to an information society. As a result companies are faced with new challenges to keep their market position, transparency and efficiency. Enterprises overcome these challenges by implementing strategies. In order to implement strategies successfully and achieve desired goals enterprises should have certain capabilities. Thus, the demand for a methodical capability management approach is growing. This chapter introduces a process for identifying, structuring, and maintaining enterprise capabilities. The guide is based on an integrated capability approach that results from a number of scientific investigations and practical experiences performed over years. Comprised of four building blocks, the capability management guide represents a flexible "engineering" approach for capability catalog developers, designers and evaluators.

12.1 Introduction

Dynamic markets, ever-shrinking product cycles and a persistent need for innovation are just a few challenges faced by companies looking for long-term success and corresponding strategies. Depending of the global changes enterprises face a variety of internal and external challenges. As a result they develop a variety of strategies, which are hierarchical arranged to reach global business goals and even pragmatic, short-term initiatives. To do so, management teams are needed to plan, transform and control organizational components like processes, technologies or

M. Wißotzki (✉) · A. Sonnenberger
University of Rostock, Rostock, Germany
e-mail: matthias.wissotzki@uni-rostock.de

© Springer International Publishing Switzerland 2016
E. El-Sheikh et al. (eds.), *Emerging Trends in the Evolution of Service-Oriented and Enterprise Architectures*, Intelligent Systems Reference Library 111,
DOI 10.1007/978-3-319-40564-3_12

219

resources. Therefore, organizations need to know appropriately about their capabilities and how to use them to handle and solve occurring challenges. What does that mean? Companies need to know themselves! In order to do so the pure identification of structural enterprise components is quickly done in most cases, but thereafter?

Our research offers a close relationship between strategic choices (e.g. projects, initiatives) and the capabilities needed for successful strategy implementation. Only then a long-term, economically efficient and structural effective existence is possible. Therefore, it is crucial to evaluate which abilities are currently available and which are required in the future, when for example:

- New business models, products or services are introduced,
- Collaborations or mergers and acquisitions are planned or received,
- New technologies or applications have to be integrated.

We understand capabilities as expressions describing factors like roles, resources, processes and information required by a company enabling the achievement of strategies. This approach is necessary to refine existing capabilities in particular models and/or to be able to merge them with existing enterprise architectures. *Why architectures?* If we look at a company as a whole there is a sum of considered components (processes, roles, departments, resources, equipment, locations) which enables business outcomes. However, a rapid and precise identification of capabilities is the basis for adequate, company-specific measures and directions of development. Decisions regarding outsourcing and insourcing, market positioning, corporate culture, innovation opportunities, etc. can be better established by capability oriented thinking. All in all, capability management needs to cope with several internal and external, market-depending and global, business and IT factors. Therefore, this section provides a guiding process focused on the following requirements:

- Scoping and preconditions for capability management
- Identification of involved stakeholders
- Identification of capability types and their relations
- Structuring of capabilities and their models as a catalog
- Governance of the resulting capability catalog

Consequently, the introduced guiding process should help to systematically derive capabilities, gathered and maintained in a repository—called capability catalog. The process is developed for all interested parties, independent of the enterprise size, branch or market. It includes working steps and specific recommended tools to visualize and notate these ones. Hence, it is adoptable to different circumstances. Generally it is addressed to all organizational departments and workers interesting in the topic itself, strategic alignment or even managing the challenges enterprises are faced to in the present days.

12.2   Strategic Management and Enterprise Architecture

The terms enterprise and organization are used in the same wording for this work. An enterprise represents an entity which is involved in a set of economic activities. It consists of a variety of subsystems influencing the entity itself—external and internal. Major external factors of influences could be triggered by markets, political, legal, demographic and/or economic. Internal factors like processes, policies, employees and culture influences an enterprise within its organization. Internal and external factors have to be considered in order to combine enterprise abilities in an optimal way, which requires a management discipline.

The origin of *management* can be traced back to the 1920s and generally implies activities like planning, organizing, staffing, directing, coordinating, reporting, decision making, budgeting and controlling. These functions are representing major management tasks excluding specific focus, instruments, methods or stakeholders. Obviously, the variety is big and primary depends on the management focus. IT and business management are examples for different focus areas that requires different management strategies as well.

In general, *strategies* could be understood as impulses for actions to be taken to reach a certain goal. The term "strategy" originally comes from the military field and represents an adjustable construct used to convert an actual state into a target state. Moreover, the own market positioning in comparison to that of competitors needs to be identified and either maintained or improved in consideration of market conditions, stakeholders, and available/required resources.

According to Fischer [12], Stutz [21], we follow five fundamental *strategic management* steps (Fig. 12.1) when it comes to the realization of business strategies to achieve defined goals.

Strategy management involves the creation of an action catalog for strategy implementation. In order to be effective, such an action catalog requires controlling techniques and a structured view of its capabilities though.

Next to the management functions above, a main challenge represents the controlling of enterprise complexity. Complexity is understood as the increasing number of elements and their relations within a system. Enterprise complexity means can be measured be relations between processes, roles, departments, resources, equipment, locations or information flows. Therefore, an enterprise-wide and integrated management system is required in order to control its complexity and

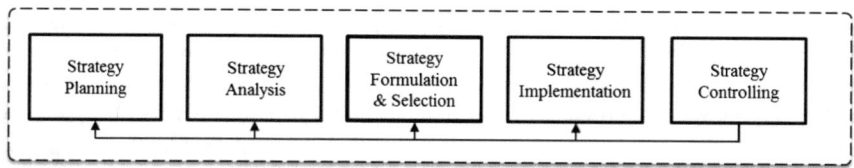

Fig. 12.1 Strategic management process (cp. [12, 21])

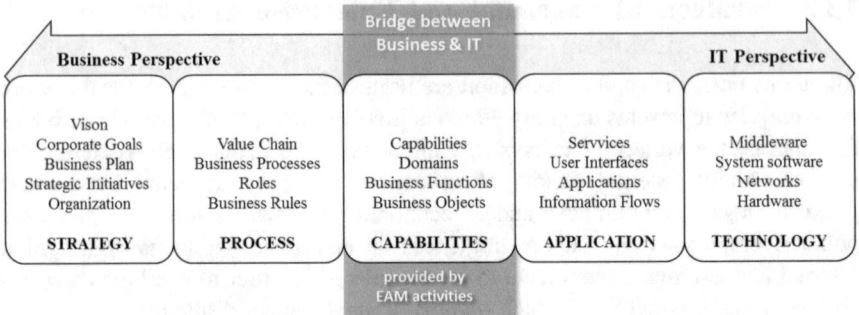

Fig. 12.2 Enterprise architecture management as mediator between business and IT

ensure organizational success. From an architectural point of view, this is supported by developing representations of enterprises in abstract architecture models.

An approach to that is called Enterprise Architecture Management (EAM) and serves as a mediator between two perspectives on an enterprise: the business- and IT perspective (Fig. 12.2). The business view is characterized by corporate goals, business model and their realization in value chains and processes considering motivational elements like corporate policies and standards, constraints, and drivers. Business goals and strategies used to achieve these corporate goals lead to the adjustment of the IT perspective, which is characterized (among others) by business information services, integration services, application landscape, middleware, networks and hardware resources.

> EAM is defines as a management practice that establishes, maintains and uses a coherent set of guidelines, architecture principles and governance regimes that provide direction an practical help in the design and development of an enterprise's architecture to achieve its vision and strategy. [2, p. 3]

Architecture Management requires to document states of the past, the present, as well as the possibility of how it could be in the future. From these states, plans are derived and transformations are triggered. The main challenge is to extrapolate which decisions are responsible for the present situation and which ones are the best to plan desired future situations. Obviously, these strategic plans should aligned with the whole enterprises architecture.

Therefore, the EAM is aligned to the strategic- and IT management. In order to align both perspectives successfully upcoming (inter) dependencies and relationships should be known and discussed. If not, problems are emphasized by the fact that business critical projects fail in 2 out of 3 enterprises which is reduced to the circumstance that a lot of decision makers failure caused by conflicting interests, insufficient information quality or decisions taken elsewhere [7]. Therefore, it is not only important to be aware of the existing challenges and problems, but also to continuously gather and asses information about organizational knowledge,

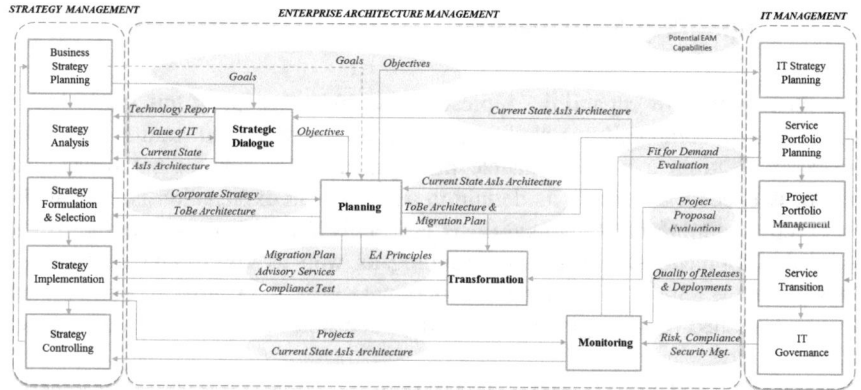

Fig. 12.3 Management perspectives, interrelations and potential capability areas (according to Fischer [12], Stutz [21], Wißotzki et al. [26])

corresponding responsibilities, available resources and processes required for the strategy implementation [27].

Exactly for this purpose we will use the capability concept, because capabilities could provide information for supporting the different management perspectives and could avoid mistakes before they arise. Each company is equipped with various capabilities that are specific to its organizations, but many of them are not aware of them.

For this purpose an elementary approach is needed that identifies capabilities required for an efficient operationalization of an enterprise strategy. Figure 12.3 summarized the three different management perspectives, its interrelations and spaces for enterprise capabilities. These capabilities should be derived systematically through structured process, gathered and controlled in a discipline called capability management.

In terms of its organizational value capabilities support:

- high-level representation of organizational activities
- strategic decisions like mergers and acquisition, out- and in-sourcing or budgeting
- transparency
- a common language between business and IT responsible
- identification of new competitive advantages
- the identification of organizational requirements for a successful strategy implementation
- scenario planning
- relating IT perspective to business value.

12.3   Capability Management

The management of capabilities represents the core concept of this work. Due to both decrease complexity as well as increase transparency, integration and (inter) operationalization, a comprehensible management approach for enterprise capabilities is needed. Therefore its definition and in-depth explanation is presented in this section. *Capability Management* is especially focused on: scoping and planning, development, structuring, maintaining as well as controlling of enterprise capabilities. It aims to optimize the economical actions in order to current capabilities, enriched by strategic and operational decision-making by the development of capabilities.

Academics and practitioners used the term "to express the ability or expertise their organization requires to engage in the execution of their business strategy and plans" [24]. In general a capabilities could be understood as the combination of resources, information, processes and business environments to reach a specific outcome (e.g. customer satisfaction, product and service quality) specified by appropriated stakeholders.

In general, they could be capabilities could be characterized as Ulrich and Smallwood [22], Bharadwaj [5]:

- intangible,
- non-redundant
- stable over time,
- process-independent, when even influenced
- hierarchical and combinable,
- heavily influenced by human resources (e.g. personnel, training),
- organizational conditionally (resulting in the ability of enterprise reconstruction),
- attributed to clear (management) responsibilities within an enterprise and
- directly market value delivering.

Thus capabilities answer the question of: "*What are we doing?*" To answer this, in most cases capabilities describe an actual summary of named expertise as well as the needed ones to fill gaps. However, the answer of "*How can we reach our goals by performing high efficient strategies?*" requires a more detailed capability concept. The concept should describe both the actual situation of required capability components, desired ones as well as procedures that closes occurring gaps. *Therefore a capability represents the ability of an enterprise to join an appropriate content depth of information and roles able to execute a specific activity with available resources under consideration of its context in order to support operationalized goals within strategies* (cp. [28]). Thus the usage of capabilities is more detailed and explicate than strategy elements, but less comprehensive than business processes [15]. Our capability model is derived from finding of Wißotzki and Sandkuhl [27], Wißotzki [27] and illustrated in Fig. 12.4.

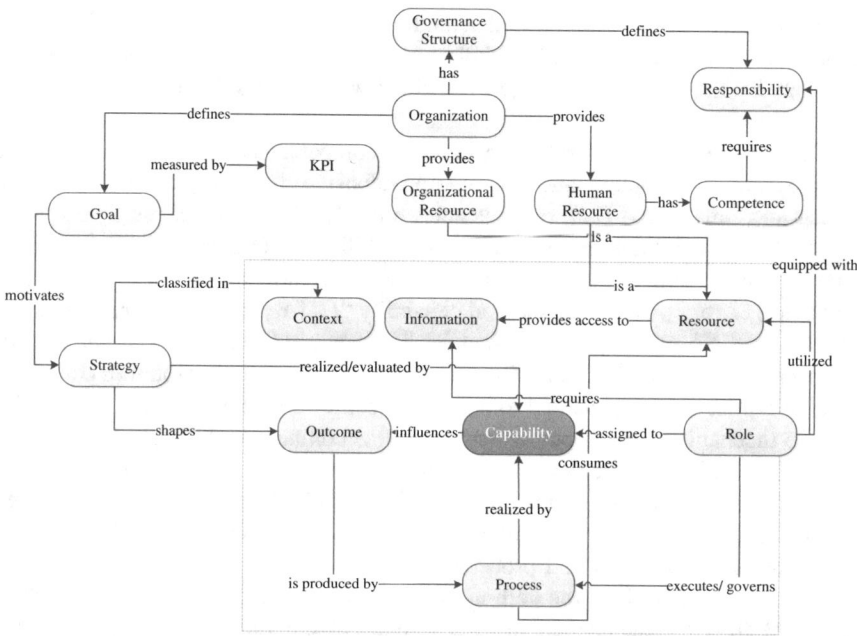

Fig. 12.4 Integrated enterprise capabilty model (IECM)

In view of the above and considering also the existence of different management contexts a wide range of *capability types* theoretically exists. Basically a distinction between external and internal focused capabilities leads to a categorization in: dynamic capabilities, business capabilities, IT capabilities, core capabilities [27].

In line with the integrated capability model (Fig. 12.4) we distinguish between business, EAM and IT capabilities characterized by its context elements, which depend on the area of application (e.g. management discipline and/or affected architecture/subject views). For instance, the context of *business capabilities* represents a combination of objects of the business architecture (e.g. product, market, or customer) and traditional management activities, whereas the *EAM capabilities* context is defined as a combination of architectural objects (e.g., application, information flow, or component) and management functions like dialog, planning, transforming and monitoring (Fig. 12.3). However "*What capability type is required for an enterprise within a certain area of application in order to achieve defined goals?*" We deal with this question using our integrated enterprise capability model again in terms of roles, resources and tasks called *descriptive elements* that starts with the definition of the already mentioned enterprise context (capability context) followed by capability' required information, roles, resources and process/activities.

Personal and organizational attitudes, skills, aims and missions, as well as their specific level of knowledge are mentioned within the capability literature [3, 10]. In order to clearly distinguish the understanding of the capability concept from concepts like competencies, abilities, skills or even processes, we start classifying the concepts by the approach of Ulrich and Smallwood [22] followed by introducing some classification criteria.

The main element to distinguish between competencies and capabilities is the individual/human context and the enterprise/business focus [28]. According to Ulrich and Smallwood [22], competencies are related to technical skill sets while capabilities and abilities are referred to social skill sets. "Organizational capabilities emerge when a company combines (and delivers on) individuals' competencies and abilities" [22, p. 2]. An organizational capability "*[...] represents an organization's underlying DNA, culture, and personality. These might include such capabilities as innovation and speed*" [22, p. 2]. Day [9] defines competencies as routines combined with enterprise investments due to activate specific functions, whereas capabilities call the mechanisms and processes creating new competencies (Table 12.1).

"Ability refers to the level of available competence, where competence is understood as talent intelligence and disposition" [11]. Both of them are addressed to reach a goal. Skills describe abilities of a person within the organization. The distinction of capabilities and processes is not favored by the variety of possible nomenclatures. Especially the verb-noun expression for capabilities (e.g. introduce products) forces misunderstanding the terms. An enterprise capability expresses "what the enterprise does" whereas a business process is about "how an enterprise operates" (cp. [8]). Processes can require granular or complex capabilities as well as the other way around. But they do not have to. One process can map different capabilities having conflicting, matching or independent requirements [15].

Next to the term differentiation by Ulrich and Smallwood [22], we demarcate the terms function, process and capability by using a set of several criteria that help us to clarify the differences (Table 12.2).

Next to the classification criteria, a set of capabilities should fulfill the following *characteristics* (independent from capability type): allows the reconstruction of enterprise organization, attributed to clear management responsibilities within the enterprise, heretical and combinable, temporally stable and process independent, but rather influenced, non-redundant and directly value delivering.

Table 12.1 Competence, ability and capability according to Ulrich and Smallwood [22]

	Individual	Organizational
Technical	Individual functional competence (e.g. expertise in marketing, finance, or manufacturing)	Organizational core competencies (e.g. financial services firm must know how to manage risk)
Social	Individual leadership ability (e.g. to communicate a vision, or to motivate people)	Organizational capabilities (e.g. innovation and speed)

Table 12.2 Classification Criteria [27]

Classification criteria	Capability	Function	Process
Decomposition	Business by strategic goals	Business by tasks/objectives	Business by activities
Extension	Enterprise wide	Unit specific	Task specific
Solidity	Enduring and stable		Change frequently
Purpose	What		How
Focus	Strategically	Tactically	Operatively
Layer	Business execution		
Modelling approaches	E.g. figures, text, Archimate 2.0	E.g. 4EM, Archimate 2.0	E.g. BPMN, EPC etc.

12.4   Capability Management Process v3.0

The initial capability management process (CMP) model version 1.0 is developed by Wißotzki [27] since 2013. The current version 3.0 is includes result from scientific investigation and practical evaluation cycles [27, 29]. This section offers a description of the CMP 3.0 (Fig. 12.5).

The process is aligned to the common accepted Business Process Modeling Notation (BPMN). Therefore, it starts with a start event and is finished after reaching the end event. Gateways are symbolized by "crossing X" and offer minimum 2 alternatives ways. For example, the transition from the 3rd to the 4th

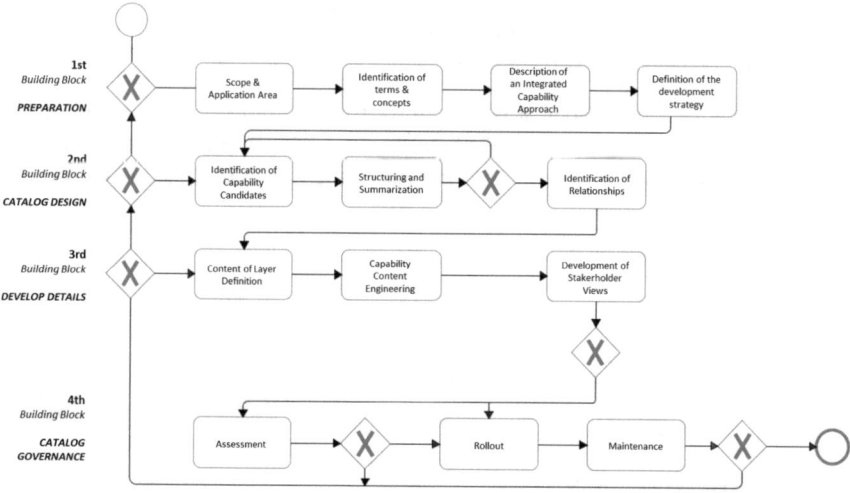

Fig. 12.5 Process model—the capability management process v.3.0

Building Block can start directly by going to Catalog Deployment by skipping the first two, optional working steps or starting right at the beginning.

The process consists of four Building Blocks (BB's) each focusing on distinct contents and having distinct outputs. In short, the first building block sets *preparation* conditions like problem, scope, and stakeholder definition. The second building block *designs* the capability catalog structure, whereas the third block *develops* the detailed capability content. The *governance* building block covers catalog evaluation and maintenance issues. Every Building Block consists of several working steps (WS). These are shortly summarized by the following central goals:

- Identification of involved parties and definition of terms and preconditions
- Identification of capability types and corresponding capabilities for operationalizing of strategic goals
- Systematic derivation of capabilities, gathered and maintained in a repository called capability catalog.

12.4.1 BB1—Preparation

The first building block defines conditions for the capability catalog to be created. Hence, the following requirements should be handled:

- Problem definition and clear scoping of the application area
- Define developer and user groups of the capability catalog
- Negotiate terms and perspectives
- Define capability types and context objects
- Agree on a common development procedure
- Form the outer frame of the catalog.

Within this Building Block it is critical to derive requirements regarding to the future capability catalog because of neglecting the current constraints. As a result, the first building block is divided into the following four, visualized working steps (Fig. 12.6):

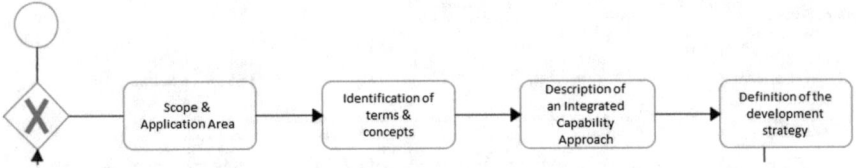

Fig. 12.6 BB1—overview working steps

12.4.1.1 WS1: Scope and Application Area

This working step forms the outer frame of the catalog but does not determine the concept of capability in depth, its level of detail, the specific context, as well as the strategy and design of the catalog. Stakeholders and the focus of the required capability model must be clarified. The involved parties have to agree on the application area and the goals of the capability catalog that is to be created. The initiators of a capability driven initiative have to agree on a capability approach due to select appropriate and specific tools and models, as well as storage media during the further development. Stakeholders are responsible to select models and tools, storage media and further approaches directly at the beginning or during the development. To summarize, the main question that has to be answered within this working step is:

For which purpose do we need capabilities?

For instance the detection of business weaknesses as well as IT alignment could be reasons for creating a capability catalog. The objectives depend on the alignment motivation like profit-, strategic- or improve-oriented motivation.

Accordingly, several driving questions are relevant for scoping:

- *What is the purpose/motivation for capability oriented thinking?*
- *Which goals and strategies need to be supported?*
- *What are the benefits for our organization?*
- *Which area of application requires a capability catalog?*
- *Are there any industry-specific capabilities that need to be considered?*
- *What is the proportion of profit about?*
- *What are the driver and constrains?*

Table 12.3 illustrates an exemplary analysis of a capability catalog's application area with respect to a potential goal to improve the business-IT-alignment.

Table 12.3 Procedure example for a goal, strategy and application area description

Goal	Improve our business-IT-alignment	*Challenge*: "IT is not able to deliver to the business strategy say 75 % of CFOs" [14]
Strategy	Development and maintenance of an architecture inventory	*Benefits*: Reliable architecture information, standardized communication, cross-company comparability of applications, reduced efforts for current landscape analysis and ad hoc reporting, ability to identify redundancies and change impacts
Application area	Enterprise architecture management	*Activities*: E.g. situation analysis, elaborate options, develop target state, road mapping and migration planning, project portfolio planning, etc.

According to human nature, there is a warily behavior towards change as long as it is not assessable. Consequently different stakeholders need to be involved by using its individual pick-up points in order to diminish this behavior and support the preparation of the capability catalog.

Which kind of stakeholders should involved e.g. managers, architects or other kind of addressee? A stakeholder analysis supports the identification of parties that are or at least should be involved, their interests, and corresponding pick-up points. Therefore, the following questions need to be answered:

- *What kind of support do stakeholders expect from a capability catalog?*
- *Who will have which benefits?*
- *Who provides the input and must be involved as a result?*
- *What is the general attitude towards the project (positive, negative, or neutral)?*
- *Who already is or needs to be informed about project goals/addressed problems?*
- *Who is essential to initiate the project and who will be affected by project outcomes?*
- *What is the general attitude of users and stakeholders towards the project?*

  - *Supporter, neutral, opponent*
  - *Interested/not interested*
  - *Engaged/not engaged*

As better the governance structure of an organization including clearly defined roles and tasks, as better works the identification of stakeholders and their agreement on development conditions. According to CEB [8] we recommend to motivate a selection of the following stakeholder groups:

- *Executive Management*: because it articulates the vision of how business capabilities will drive enterprise value, ensure that the senior management are engaged in the initiative, approved the overall concept and release the budget for the catalog development.
- *Senior Management*: because it represents knowledge carrier of organization's mission, operations, and performance objectives, identifies potential capability stewards who will be accountable for required information, validates drafted capabilities to ensure that they accurately represent activities of their business unit, function or organization.
- *Middle Management*: because its articulate how their units operational activities link to strategic goals, can validate drafted business capabilities to ensure that they accurately reflect the activities of their function.
- *Enterprise Architects*: because its knowledge about enterprise architecture, modeling techniques and industry frameworks is essential to engineer capabilities in depth.
- *Business Architects*: is deep understanding about most important business activities is crucial for capability engineering.

We recommend three facts that should be taken into account when selecting stakeholders:

1. *Find stakeholders who could give feedback*: Comments, changes, additions, incomes and outcomes to a capability catalog project like:

   (a) Managers with an enterprise-wide understanding.
   (b) Domain experts with knowledge on what they do.

2. *Integrate a business unit or functional division*: Who is interested in future changes and has a positive attitude towards the topic. For example: Development of a capability catalog to support the relations between business and IT.

3. *Locate a executive or senior manager* by illustrating benefits of a capability oriented thinking.

Most work recommends the senior and middle management as main addressees. These hierarchical layers provide the majority of required information and resource responsibility for capability engineering initiatives. The definition of the general scope and application area, overall budget and distribution of resource responsibilities is embedded in executive layer. Architects combine the different (architectural) views with the management information. If for any reason a distinction in such management groups not possible chosen stakeholders should provide at least the following characteristics, rated from low to high importance:

- Architecture work and technical mastery like technical and deep domain expert knowledge
- Strategic thinking like long-term business oriented thinking
- Business and IT understanding like knowledge about enterprise processes and architecture
- Engagement skills like collaboration, decision-making, change management, facilitation

For all involved stakeholder a common (moderate) business language as well as specific enterprise vocabulary should be in order to document results understandable, transparent and identifiable. Especially, capabilities should be defined and documented with understandable terms, which at least depend on involved stakeholders and its chosen languages. We recommend that any form of documentation should be written from an outside-in-perspective to allow addressees and may project external stakeholders to understand it.

Documentation Example: Relationships and dependencies of this WS could be documented within a project concept or scoping description, visualized by models like goal models, business models and/or stakeholder diagrams (Fig. 12.7), bubble diagrams, portfolio charts, organizational charts, spider charts etc. However, storing of reached results is of significant importance and could be supported by (1) centralized (2) access to.

Fig. 12.7 Visualization
example—stakeholder
diagram

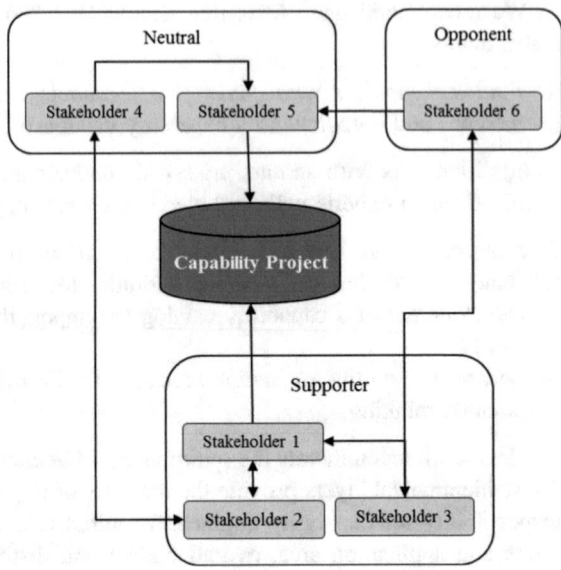

1. *Centralized* and/or distributed data storage; e.g. MS SharePoint Services, local
   and/or global storage server and file systems, internet based services
2. *Access* to documents and database(s): e.g. provided servers, online-platforms,
   wikis, personal (electronic) files, cloud services, knowledge management
   systems

   Finally, Table 12.4 summarized the inputs, throughputs and outputs of this WS.

12.4.1.2   WS2: Identification of Terms and Concepts

The understanding and choice of a capability concept may vary among relevant
stakeholders. So a common understanding between different stakeholders with
differing languages (not in the understanding of spoken language, but rather the
specific working vocabulary) must be found. Starting with a general capability
approach may create a common understanding of the perspective at hand.
Nevertheless, obtaining an overview of already existing definitions and concepts is
advisable in order to either use or extend present concepts. At this point, the global
requirements of the capability catalog development are defined and the existing

Table 12.4 BB1.WS1—Summary

Input	Throughput	Output
Vision, strategy and goals, business model, organizational model	Scoping and basic conditions for the capability catalog	Approved scope and stakeholder commitment

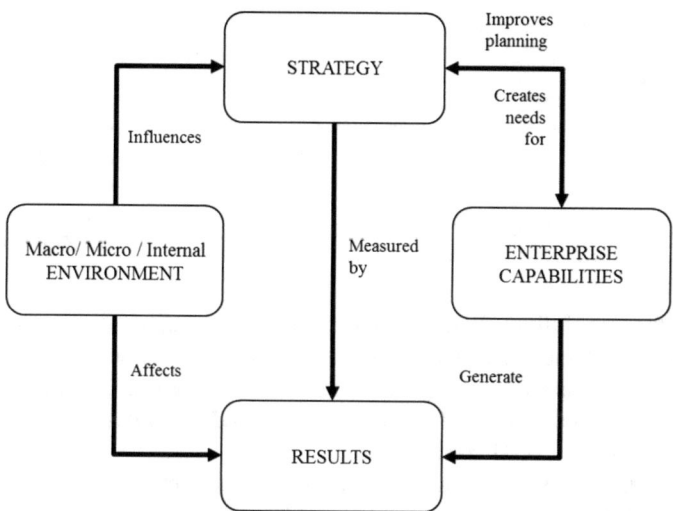

Fig. 12.8 Documentation example to document capability relations within an enterprise (according to Greski [15])

concepts are compared and enhanced by missing components. This working step identifies terms and perspectives to define a consistent capability concept.

How do we extend already existing, documented terms and concepts?

We recommend a deductive procedure: starting with a general example of the capability approach for a common understanding, specific constellations and elements can be derived during this WS. The brief overview of existing and documented terms must be extended and modified in detail. For example, present approaches can be adapted to new situations or its relation within an enterprise can be reconsidered (Fig. 12.8).

The driving questions are:

- *Are there any existing capability definitions, maps, projects, catalogs, contracts etc. within the organization?*
- *What is the understanding of used terms about?*

   – *Internal, external, common*

- *What hopes are desired by developing such a catalog?*
- *How is the concept of capabilities applied?*
- *How can we implement the identified capability concept into the existing organizational structure?*
- *Which degree of detail is currently reached?*
- *Which architecture types are involved and/or influenced?*

Table 12.5 BB1.WS2—Summary

Input	Throughput	Output
Scope and application area	Identifies used terms and perspectives to define a consistent capability concept	Approved capability working definition
		Approved architecture concept
		Idea of required capability types

Enterprise capabilities influence the results of transformations specified and measured by strategies. Architectural components like processes, information, roles and physical resources are assigned to these enterprise capabilities. These basic components are part of the business architecture model and has to be considered in order to analyze its causal correlations.

Documentation Example—Results like written explanations of the single terms and statements are sufficient and should be collected in a central glossary, (internal) wiki and/or file repository (see WS1). Relationships between defined concepts could additionally be visualized within informal or formal models. Nevertheless, no single visualization tool or technique could be recommended for this working step, but we recommend to orient themselves on enterprise standards like modeling languages, or knowledge management procedures. Finally, Table 12.5 sums up the inputs, throughputs and outputs of this WS.

12.4.1.3 WS3: Description of an Integrated Capability Approach

In this step, the fundamental capability approach and context definition is worked out. According to Abowd et al. [1], a context describes any information that can be used to characterize the situation of an entity. As mentioned indicated, an integrated capability approach represents an object-based concepts including its context and relations within the enterprise architecture. Referring to Buckl et al. [6], capabilities have either a direct or indirect relationship to (other) architectural objects. The enterprise context contains the description of any information characterizing a specific situation. Consequently, the sum of internal and external factors influencing a specific situation has to be identified and defined. Within this working step these architectural objects including its relations and enterprise context are assigned to a capability concept in order to specify its structure and type.

What is the capability context like?

Our Integrated Enterprise Capability Model (IECM) supports the identification of specific capability types required for effective operationalization of specific goals and strategies. In line with an enterprise architecture approach (Fig. 12.9), both the application area and the elements required for a capability could be identified, assigned within an EA and finally results in a specific capability type. We distinguish between

Fig. 12.9 Capability types
and context relations

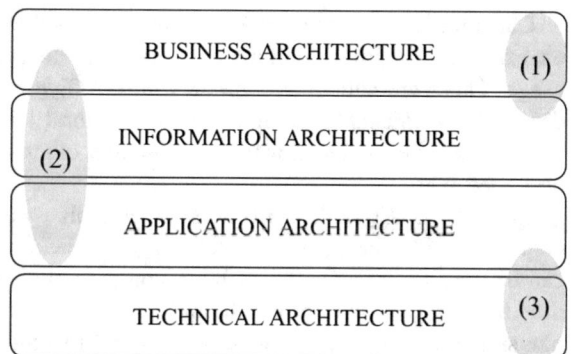

three basic capability types: (1) Business Capabilities $\rightarrow$ Business context, (2) EAM
Capabilities $\rightarrow$ architectural mediator context, (3) IT Capabilities $\rightarrow$ IT context.

Each type is imbedded in an enterprise context, which in turn depend on the area
of application (BB1.WS1). For instance, the context of *business capabilities* represents a combination of objects of the business architecture (e.g. product, market,
or customer) and management activities, whereas the *EAM capability* context is
defined as a combination of architectural objects (e.g., application, information
flow, or component) and management functions.

> What capability characteristics are required for an enterprise within a context in order to
> achieve defined goals?

We deal with this question using our integrated capability approach. Next to the
enterprise context, the specific definition of a capability requires an additional set of
elements: the required information, roles/actors with competences to help create a
specific outcome, the relevant activities or processes, and appropriate resources
(Fig. 12.10).

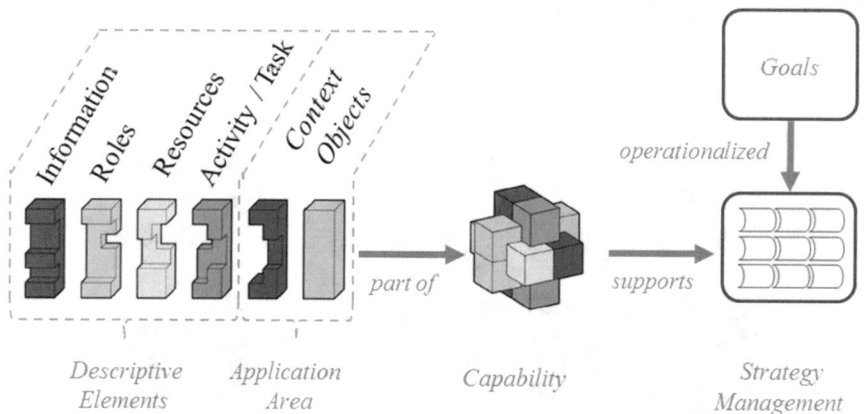

Fig. 12.10 Conceptual structure of a capability (cp. [27])

Capabilities are connected to an overarching subject (e.g. application area) or an environment (internal/external) and describe the specific situation of capability usage. These are called "*Enterprise Context Objects*". The capability context and descriptive elements are assigned to architectural layers of the organization's EA, which could be broken down more detailed other than the conceptual examples described here.

Procedural Example—Business objects like costumer, contract, and order combined with management functions (e.g. planning, implementing, controlling) can be used for a business capability context definition (Fig. 12.11). Furthermore, time horizon (e.g. current and future) and/or more specific management functions like situation analysis, elaborate options and roadmaps could be used.

Procedural Example—For an EAM capability we refer to architectural layers and/or architectural objects like application or information flow. Furthermore, we recommend to use specific EAM management function to define an EAM capability context (Fig. 12.12).

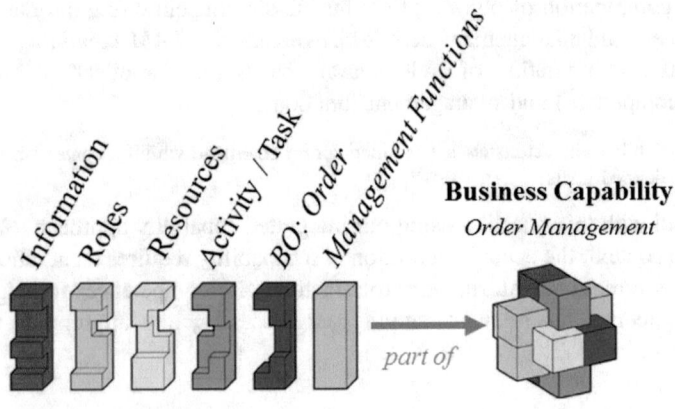

Fig. 12.11 Example of a business capability

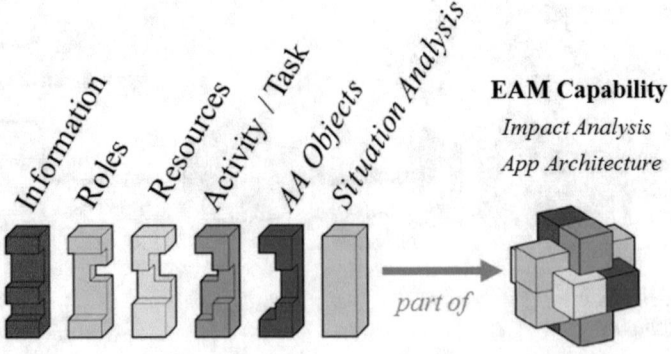

Fig. 12.12 Example of a EAM capability

Table 12.6 BB1.WS3—Summary

Input	Throughput	Output
Capability concept	Description of the specific capability type by context definition	Integrated capability approach

The indicative questions, helping to define which context objects are required:

- *Do we need related, underlying and/or linked information, roles, resources, and processes for our capability definition?*
- *Which descriptive elements are important for us?*
- *Which capabilities do we need and use?*
- *Which capability types exist in specified practice?*
- *Are there any context objects derivable from the application area? If yes: How?*

Finally, Table 12.6 summarized the inputs, throughputs and outputs of this WS.

12.4.1.4   WS4: Definition of the Development Strategy

Questions of how the catalog is constructed are answered in this section. During the development of strategies, it is necessary to obtain management approvals and support. In addition, all relevant organizational units and employees (*BB1.WS1*) have to get access to required information and documents. In fact, informing relevant stakeholders about, e.g., the upcoming activities and the corresponding timeframe is essential in order to obtain the required support. The relevance of the overall project to the enterprise, the purpose of the capability catalog, a time schedule, planned activities, the involved parties, a common understanding of how capabilities will be applied—all of these aspects need to be clear and/or available right at the beginning. The main objective here is to create openness among the involved parties or, say, stakeholders to upcoming analyses in order to have a positive influence on both quality and correctness of the identified capabilities. The need for personnel and monetary resources required in the context of a capability development project may have to be justified during the first building block as well. The following aspects may generally support the value justification:

- Added value of the capability catalog in accordance with the overall performance of an enterprise, e.g., cost savings or quality enhancements
- Development of competitive advantages with the aid of capability-based planning and investment
- Improvement of the documentation and auditability of organizational requirements used to achieve goals

What is the overall capability development strategy?

Two situations can be differed and should be considered during the definition process of this WS: there is already is an existing catalog or a new catalog has to be developed. Furthermore, the strategy definition should include: purpose, time schedule, planned activities, stakeholders, resources, common wording and understanding, documentation and engineering approach. Obviously, the previous three working steps provide the basis for the development strategy, e.g. the purpose and addressees defined in *BB1.WS1*, the defined elements like resources in *BB1. WS2* as well as the context (*BB1.WS3*) to derive possible effects and, as a result, to plan comfortable in time in this working step.

Procedural Example—Justification of strategy alignment can be reached by:

- *Increasing competitive advantages* with the aid of capability-based planning and investment
- *Added value* of the capability catalog in accordance with the overall enterprise performance, e.g. cost saving
- Improvement of communication and documentation purposes
- Top-down planning procedure—starting with basic goals and directions to specific objectives and required activities.

The driving questions are:

- *How can we anchor strategic goals into our capability catalog?*
- *How can we learn and translate existing processes into capabilities?*
- *What personnel and financial resources are needed to realize the development project?*
- *How can the output of the project be valued and accordingly measured (financial, organizational, personnel)?*
- *How are the capabilities used or usable in practice?*
- *How is the timeline and division of responsibilities for each activity?*

Procedural Example—a comprehensive strategy planning view should consider the following three levels (cp. [22]):

- *1st Intellectual Level*: It has to make sure that stakeholders from top to bottom know what the strategy is, what it is crucial influenced by and what is its need and importance.
- *2nd Behavioral Level*: Time plans and the real spending in strategic issues as well as their degree of influencing must be analyzed.
- *3rd Procedural Level*: The continual invest in strategic essential procedures must be stressed.

Next to the planning perspective, the engineering approach, the modeling languages and modeling/documentation software tools have to be defined. For the engineering approach we recommend three different ways (according to Espana et al. [11]):

- *Goals-oriented*: Starting with defining and modeling a goal hierarchy, required capabilities to reach the organizational intentions and objectives must be analyzed. Top-down-modeling is recommended.
- *Process-oriented*: Starting point is a process underlying a business model, which is further modeled and defined in order to adopt it in different scenarios. This approach assumes (at least) an existing organizational process model.
- *Concept-oriented*: Static aspects (e.g. structures, materials, customer profiles) of enterprises, called concepts, have to be modeled and analyzed in order to illustrate organizational knowledge.

The selected engineering approach directly affects the modeling languages, provided syntax, visualization techniques and at least the software tool decision. The authors recommend a central storage/repository idea for the capability catalog documentation due to more comfortable and easy data access by other information systems. Furthermore, isolated solutions and redundancies are avoided as far as possible. An overview of possible different engineering approaches is summarized by Table 12.7 (cp. [11].

Table 12.8 sums up the key points of this WS.

Table 12.7 Capability engineering approaches (cp. [11]

Aspect of comparison	Goals-oriented	Process-oriented	Concept-oriented
Starting point of modeling	Goals	Process(es)	Static aspects
Basic intention	Enterprises gain to reach their goals. Capabilities fulfill them	Capabilities are a set of processes	Any kind of resources flow into capabilities as well their effects
Preconditions with respect to models	Goal hierarchy	Process model	Structured and defined organization (worker, organizational structure, resources)
Primary stakeholders	Executive and senior mgt.	Domain experts, product owner, senior and middle mgt.	Product managers
Degree of flexibility of the modelling strategy	Iterative and incremental modeling process	Flexible process engineering with regard of capability design revision and cope with ill specified goal or concept models	Flexible with regard to the business process specification and cope with different levels of concept granularity
Organizational impact	Reinforces strategic vision and clarifies the IT-business alignment	Improvement of the total enterprise context	Grouping of organizational concepts

Table 12.8 BB1.WS3—Summary

Input	Throughput	Output
Integrated capability approach	Detailed scoping of the project and alignment to the outcomes of the previous working steps	Definition of the project plan and engineering approach → development strategy

12.4.2 Catalog Design

Subsequent to the determination of the basic conditions within the first building block, the design of the capability catalog is initiated. Hence, capability candidates are identified, collected, structured as well as their relationships identified. The building block consists of the following three working steps (Fig. 12.13):

According to Ulrich and Rosen [23], Wißotzki and Sandkuhl [27], the following list summarized a number of basic principles important for capability identification and definition:

- BB1.WS1: Application Area affects capability context and its type.
- BB2.WS2: Capabilities define what is done, not how to do something.
- BB1.WS2: Current capability understanding.
- BB1.WS3: A capability is defined by its set of descriptive elements.
- BB1.WS2: Capabilities are nouns.
- BB2.WS1: Capabilities are defined in the business language of its application area (i.e., there should be no technical terms for describing business capabilities).
- BB2.WS2: A capability should be enduring and stable, not volatile.
- BB2.WS2: Capabilities are not redundant.
- BB2.WS2: There is one capability map for an application area.
- BB2.WS3: Capabilities can have relationships to other capability types.

12.4.2.1 Working Step 1: Identification of Capability Candidates

The phase starts off with the "capability candidate identification." The focus of this activity is the definition of the first capabilities. Prior to any analyses, it is important to accurately define the area of application and coordinate the required work (*BB1.*

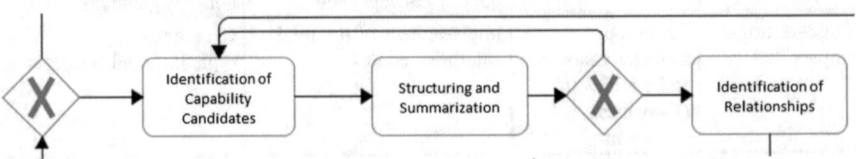

Fig. 12.13 BB2—overview working steps

WS1). The area of application determines the content and concepts that are significant for the identification process. Therefore, the output of *BB1* provides the basis for the planning of required identification activities, involved experts, and the effort estimation. For the actual identification process, there are several possibilities that have been successfully used in other fields such as enterprise modeling. According to Wißotzki [27] different identification methods with respect to their field of application exist e.g. brainstorming, survey, document analysis, written cases or moderated workshops. Initial activities for identifying capabilities should be kept as short as possible. In general, these initial activities result in a roughly structured collection of individual capabilities or at least capability ideas.

Procedural Example—The following Table 12.9 illustrates a couple of examples of typical industry-related business capabilities in order to provide guidance for a simple one dimensional (only one capability context element) capability identification.

Next to the one dimensional example above, we recommend an additional approach for a multidimensional capability identification process with two or three capability context elements. The origin of this identification concept is a so-called *"capability identification matrix/cube."* At the axis of the identification matrix the context elements of a capability type are positioned. In case that more than two context elements are defined, the set has to be stretched to multidimensional spaces. Engineers should consider that this set is much more complex. Consequently, we recommend not more than three context elements in order to avoid complexity at the beginning and enhance handling and understanding, because the identification of first capability candidates represents the main objective of this working step.

Example—In context of our IEC approach an EAM capability like *"Impact Analysis Application Landscape"* (IAAL) could be identified as follows. The already defend context elements "architectural objects" and "management functions" name the X- and Y-axis of the capability identification matrix. At the X-axis, we position the following simplified EA management processes:

- *"planning"*—involving the phases: situation analysis, elaborate options, develop target state, road mapping and migration planning, project portfolio planning.
- *"transformation"*—involving phases: project set-up, design solution, implement solution, roll-out.

Table 12.9 BB1.WS3—Summary

Capability context—industry	Business capability examples
Utility	Claims management, network capacity management
Automotive	Production equipment manufacturing, supply chain management
Banking	Safety management, credit management, compliance management
Software	Product life-cycle management, test and validation management
Mining	Production planning, core extraction, waste management

- *"monitoring"*—involving phases: control and evaluate EA, manage change needs.

The Y-axis (architectural objects) contains architecture elements of the business, application and technical layer (Fig. 12.9). In order to provide examples for upcoming explanations we exemplary assign a set of elements to the defined architectural layers illustrated in Fig. 12.14.

For our example an application architecture objects called "application" is selected. This architecture element represents an IT system that provides features in a business manner to user or other applications. Applications based and operate on elements of the technical architecture. The matrix cell at the intersection of the "application" object and for this example "situation analysis" of the "planning" phase represents a possible EAM capability. The engineering team has to discuss and decide about the meaningfulness of this intersection in terms of "Does this intersection represents a capability for us?" The current situation of an application and its relations to other architecture objects provides important inputs for additional process like transformation or changing activities. Consequently, the question

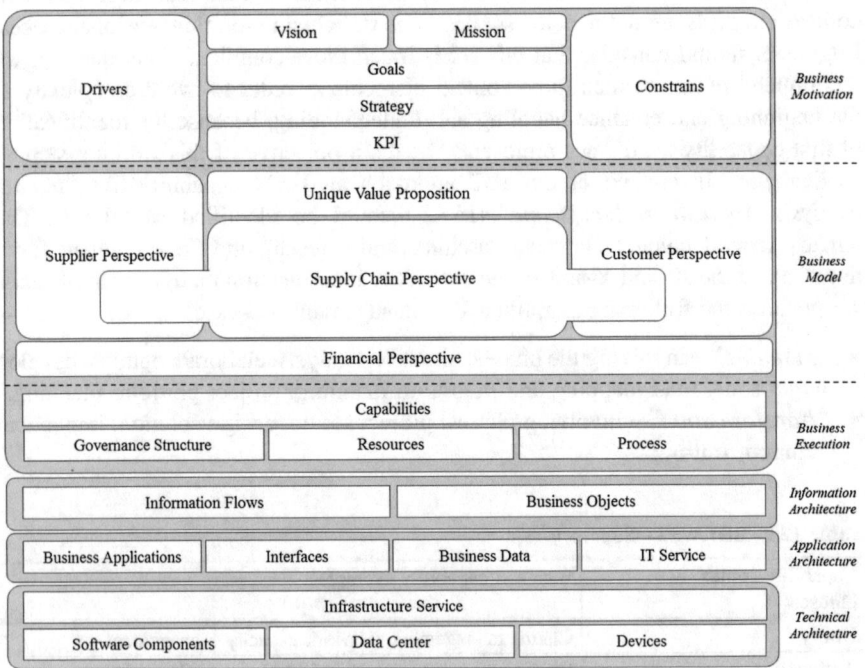

Fig. 12.14 Example for an enterprise architecture and typical layer elements (cp. [20])

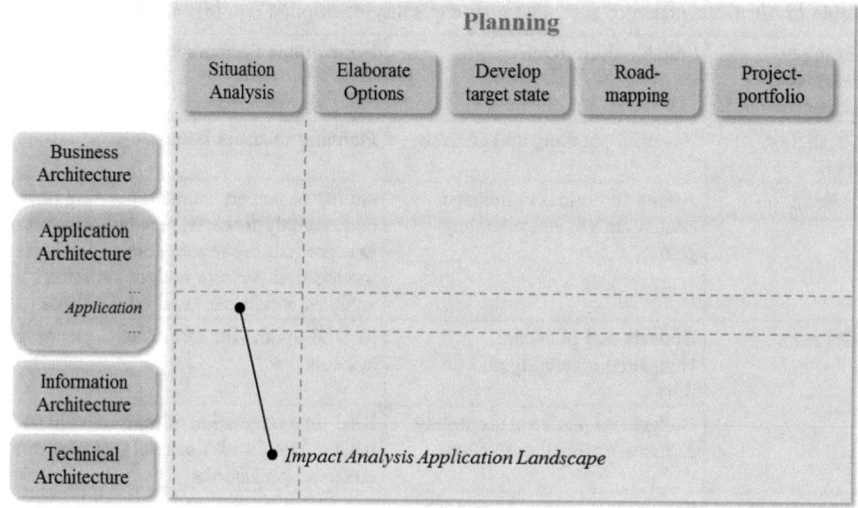

Fig. 12.15 Concept of an EAM capability identification matrix—management context element planning

above could be answered with yes. The upcoming EAM capability is exemplary called "*Impact Analysis Application Landscape*" capability. Figure 12.15 illustrates the example.

Naming of capability represents another important issue of this WS. In this context two options for the nomenclature of capabilities could be distinguish:

1. Noun declarations
2. Verb-noun declarations

In terms of classification purposes (cp. Table 12.2) we recommend to name capabilities by nouns, whereas other organizational elements (e.g. processes, business functions, value streams) should use noun-verb declarations. A suitable declaration facilitates a fundamental goal of a capability management in terms of being an instrument of communication between different enterprise perspective (Fig. 12.2) by enhancing the understanding and transparency of *what* these perspectives do. Nevertheless, even at this early stage suitable declarations should fulfill the following definitions:

- As simple and short as possible,
- Conclusive and consistent,
- Focused and transparent,
- Describing and comprehensive,
- As significant as possible,
- Statement-like.

Table 12.10 Comparison of good and poor capability description (cp. [8], p. 17)

Example business capability	Suitable declaration	Unfavorable declaration
Capability name	Financial planning and analysis	Planning financial issues
Description	Ability to build and manage annual budget and operating plan	Ability to prepare annual operating plans and monthly forecast, generate product line profitability reports, create financial models and perform budget variance analysis within the required timelines
Outcome	Budgets and plans are completed efficiently and on time	99.9 % uptime for all financial planning systems
	Budget and plans are accurately estimate financial outcomes	Real-time integration of transactional and master data for all financial planning and analysis applications
Responsible	Senior mgt., corporate finance	–

Table 12.11 BB2.WS1—Summary

Input	Throughput	Output
Comprehensive development strategy	Identification of capability candidates	Roughly structured collection of capabilities

Table 12.10 represents an example for a suitable and unfavorable capability declaration.

Questions helping to identify capabilities are:

- *What kind of abilities do we need to do our business?*
- *Could we derive capabilities from core processes?*
- *Could we derive capabilities from value chain?*
- *Could we derive capabilities from business functions?*
- *What kind of analysis methods could we use for an initial identification?*

The identification should be made by a core team, which is enlarged by domain experts during the next working steps or further iterations. These domain experts obtain special knowledge from practice or scientist in order to provide completeness and different perspective regarding a capability. However, this WS ends in a roughly structured collection of capabilities that could be visualized by different techniques.

Documentation Examples—Cluster maps (box-in-box), Capability Identification Matrix (Table 12.11), mind maps, simple lists or text or other collection documentations are usable. A common capability visualization technique repents the

Fig. 12.16 Visualization of a business capability collection in a cluster map

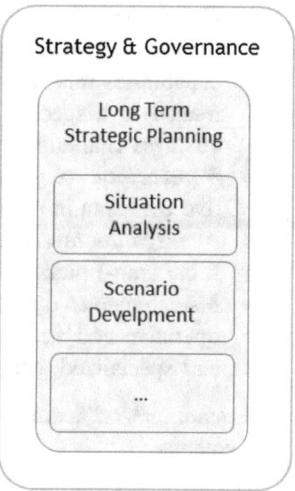

cluster map (Fig. 12.16). However, irrespective of the chosen techniques each of them should be centrally saved as editable document for the engineering team and authorized stakeholders.

12.4.2.2   Working Step 2: Structuring and Summarization

Within the step "structuring and combining," redundant elements are removed and capabilities that have a strong coherence as to content are aggregated or further specified. Content-related aspects are combined to create a catalog that is both easy and clear to understand. Subsequent to first refinements of the capability catalog, participants work on additional iterations with the aid of the collected questions and critical comments in order to suggest further changes and enhancements.

The objective of this step is to classify identified capabilities, create a consistent structure, fix capability names and prepare stable descriptions in order to keep the amount of as small as possible, but as large as needed. Therefore, the following activities should be considered:

- Review of the first substantial results of the brainstorming activities
- Pooling of redundant elements with similar stating points
- Register coherences between capabilities (aggregates, interrelation)
- Further analysis and reorganizations are needed.

Consequently, initial identified capability candidates have to be analyzed, discussed and, where necessary, restructured. Restructuring can be differed into:

1. *Removing*—of unnecessary elements
2. *Grouping*—similar capabilities are either pooled or integrated using appropriate criteria like:

- *Collection criteria* e.g. all capabilities of the same business, all capabilities required for value proposition; capabilities required to reach defined results, capabilities related to specific roles, task or business functions; capabilities related to a specific business partner, capabilities required to overcome a business challenge.
- *Nomenclature* e.g. nouns for capabilities and noun-verbs for other descriptive elements in order to differentiate them at first glance.
- *Aggregation levels* e.g. high levels for a first complete overview, pooling of same (sub-) hierarchies levels.
- *Miscellaneous* e.g. competitive and support, importance, customer faced, operative and strategic, business and IT, available and theoretical, general and specialized, enabling and disable.

3. *Extending/Modification*—includes the further specification or aggregation of elements.

The following driving questions support the structuring process:

- *Are there similar or rather redundant capabilities? Yes: Is it possible to aggregate or reduce these ones? Or Do they have to be more specified in relation to a better distinguishing?*
- *Is the capability catalog unambiguous and easy to understand by the stakeholders? No: Are there any techniques like reduction, composition and decomposition to increase it?*

After reducing and summarizing, content-related capabilities they can be restructured, grouped or aggregated like illustrated in Figs. 12.17, 12.18 and 19.

Example—The business capability decomposition of strategy and governance management can look as the following (Fig. 12.19).

Summarizations and new structures should be accepted by the involved parties, especially if capabilities are removed or modified to answer questions like: Does our new catalog structure represent our application area? There are two possibilities to answer the question above:

1. Yes: Everything is fine. The structuring process is finished. The next working step can be started.
2. No: More questions have to be asked and answered:

   (1) *How can we reach more acceptances?*
   (2) *Are there serious reasons for resistance?*

Questions and critical comments have to be documented in formal and informal. Moreover, in the course of several iterations, it is necessary to use suitable documents in order to implement a resistant and stable documentation process.

Especially from here, we recommend a combination of the capability identification matrix (Fig. 12.20) and a software tool. Next to identification purposes the matrix concept provides a structuring concept for this stage (Table 12.12).

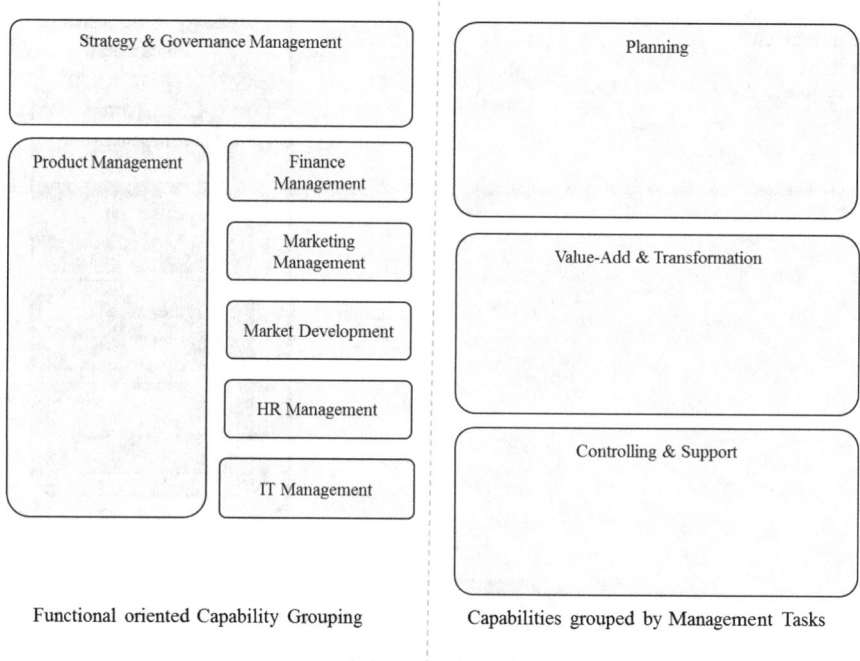

Fig. 12.17 Restructuring example—capability grouping

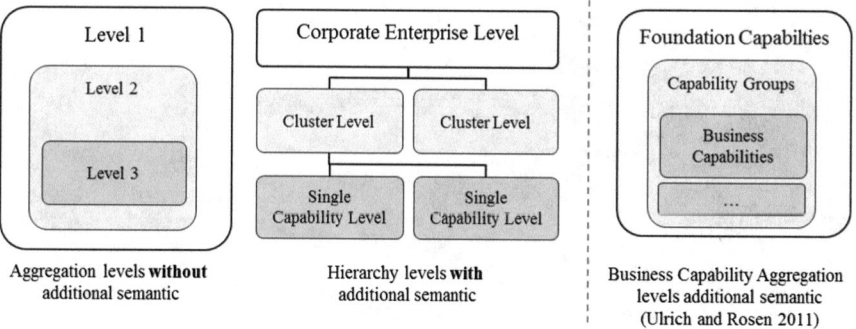

Fig. 12.18 Examples capability aggregation levels

12.4.2.3   Working Step 3: Identification of Relationships

Since the collected improvement suggestions just provide content-related horizontal breadth and not particularly vertical a depth of a capability catalog, it is necessary to conduct further analyses and reorganizations. In addition to an improved level of detail that is achieved in *BB3*, dependencies among capabilities need to be

Fig. 12.19 Capability levels
of strategy and governance
management

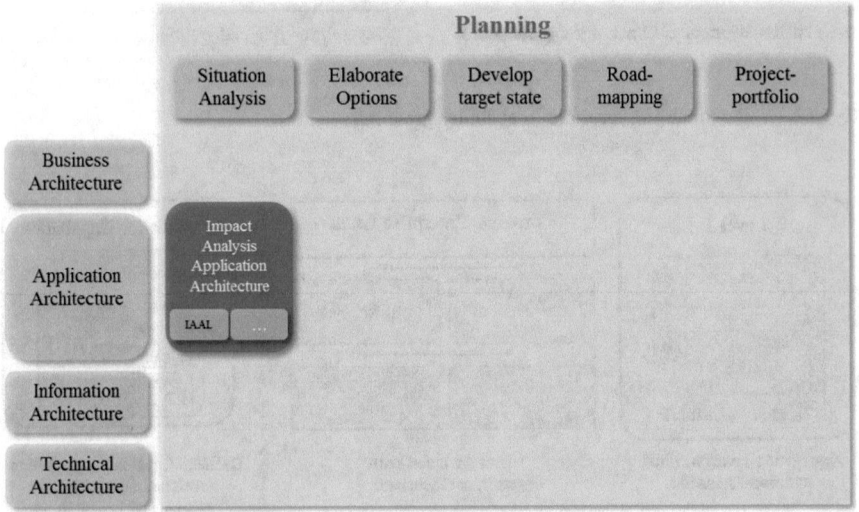

Fig. 12.20 Documentation example for an aggregation of the impact analysis application architecture

Table 12.12 BB2.WS2—Summary

Input	Throughput	Output
Roughly structured collection of capabilities	Classify capabilities, create a consistent structure, and fix capability names and prepare stable descriptions	Capability catalog structure

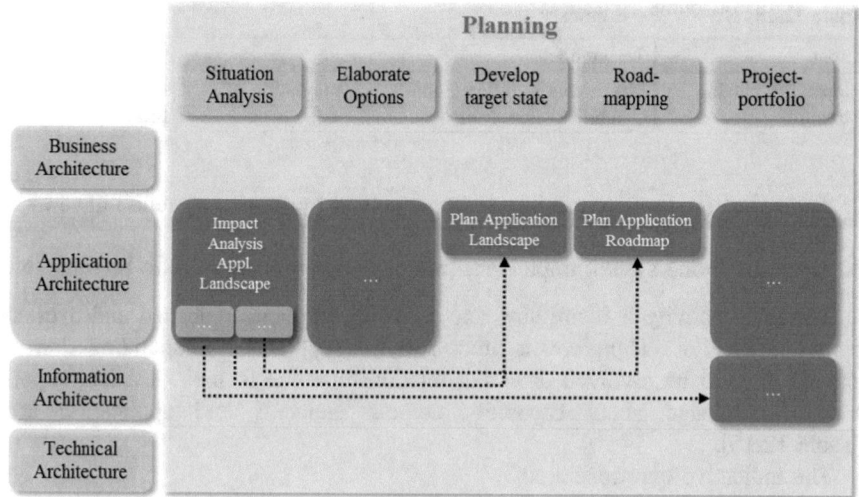

Fig. 12.21 Visualization example for relationships in the capability identification matrix

identified and documented previously. During the step "relationships identification," different relationships are documented and analyzed. As a result of identifying missing relationships, removing inconsistencies, and discovering gaps, there is an enhancement of both the knowledge represented by the catalog and the understanding of capabilities being available within an enterprise (Fig. 12.21).

The different types of relationships have to be documented and analyzed. Basically, it can be distinguished between the following relationship types:

- *Dependencies and correlations*—One capability needs another one. Informative dependencies are a subtype in term of information need.
- *Interdependencies*—Mutual reliance between (at least) two capabilities.
- *Independencies*—Capabilities exist side by side without any link.
- *Synergies*—The sum of capabilities has more value than the separate ones. The entire relation of them is in the interest.

The following subtypes of relationships are rated as useful within a capability catalog:

- *Informative*—One capability needs information from other ones.
- *Supportive*—One capability is a precondition for other ones.
- *Functional*—Two capabilities represent different connections/aspects in the same character (one of the context elements are identical).

There are three main tasks fixing relationships:

1. *Find missing relationships*: gaps must be discovered and missing relationships identified and inserted

Table 12.13 BB2.WS3—Summary

Input	Throughput	Output
Capability catalog structure	Identification, differentiation and integration of capability relationships	Capability catalog v1.0

2. *Redundancies* (=one capability is implemented by several systems) have to be removed
3. Overlaps (=one system implements multiple capabilities) have to be removed

The main activity is identifying and adjusting implicit, undesired and overlapping relationships. Therefore, a process (e.g. [13]) and corresponding domain experts have to be involved detecting relationships due to their practical experiences, knowledge about capability context elements and application area (Table 12.13).

The indicative questions are:

- *What kinds of relationships exist between the capabilities?*

  – *Informative, supportive, functional*

- *Are the relationships totally identified and defined?*

  – *No: Where are the missing ones? How can we fill the gaps?*

- *Are there any wasted, inconsistent or unnecessary relationships?*

  – *Yes: How can we eliminate them?*

- *Do the stakeholders agree to the identified relationships?*

12.4.3 Detail Development

As described, capability management is typically an iterative process of identifying, defining, controlling and maintaining. Thus, it is completed once when every capability is described in a sufficient level of detail for supporting the specific strategy implementation of an enterprise. The third building block is responsible for the refinement and renewing of already achieved results by applying the following steps (Fig. 12.22):

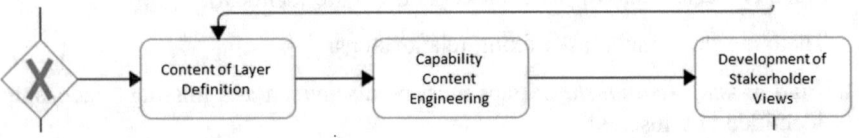

Fig. 12.22 BB3—overview working steps

12.4.3.1   Working Step 1: Definition of Content Layer

The initial step of the third building block, "catalog content layer definition," addresses the definition of the content and associated depth in order to provide both a final structure and relations of the capability catalog details. This step is important in case the catalog needs to achieve a high level of detail in terms of content (e.g., by specifying descriptive elements and defining assessment criteria). Content layers are crucial to define content in an associated depth in order to provide both a final structure and an order of the catalog. Therefore, the descriptive elements, the content objects and other needed terms have to be specified in a high level of detail.

The example in Fig. 12.23 illustrates a three-level approach for the content layer definition. The capability identification matrix represents the first level and is used to identify contextual capabilities. At the content level the descriptive elements are precisely specified. Last but not least, different kinds of assessment criteria and procedures are defined at the third level.

Procedural Example—Weldon and Burton [24] differ between the following three layers due to being familiar, logical, comprehensive and adaptable to all stakeholders:

- Level 0: *Contextual*—identification and naming of context objects (e.g. sell, market, service, partner, procure)
- Level 1: *Conceptual*—identification and naming of related capabilities (e.g. for the context object "sell" there might be "sell miles to partners" and "sell miles to members")
- Level 2: *Logical*—further and more detailed sub-classification by chosen criteria (e.g. domestically sell or export for partners or members)

In particular the middle management might need more details in order to elaborate planning scenarios like level 1–2 from above, whereas the executive

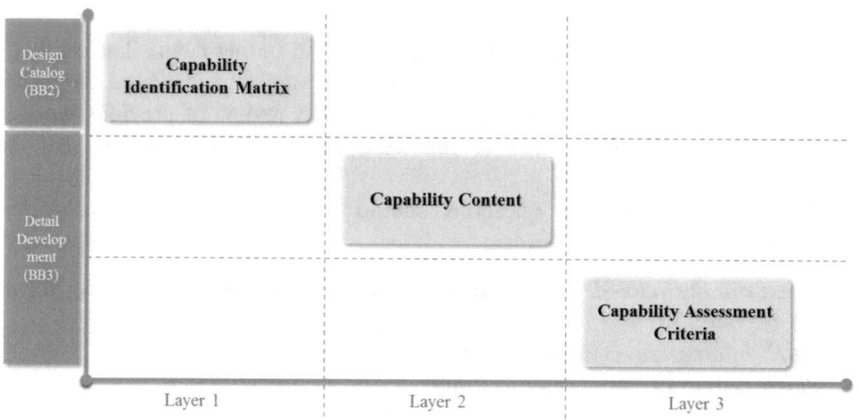

Fig. 12.23 Example for content layers [28]

management is more focused on high-level capability content in order to receive global and complete overviews like level 0–1. First content layer depth impressions can be derived from assigned aggregation levels of BB2.WS2.

Questions, helping to identify and define content layers are:

- *Which degree of detail should be reached/is required?*
- *Which layer differentiation should be used?*

  - *Foundation, groups, types, descriptive elements, KPIs*
  - *Contextual, conceptual, logical*
  - *Corporate, cluster, single capability, descriptive elements, KPIs*

- *How many content layers are practical and rather needed?*
- *How are the single layers connected?*
- *Which issues should be described in each layer?*

Easy, clear and appropriate vocabulary is recommended to define the layer names and its content requirements. Due to the increasing content complexity, the more substantive documentation should be made within a software tool. Hence, both the previously identified relationships between capabilities as well as between descriptive elements can be taken into account. Regarding the levels, different approaches clarify the terms, needs and visualizing models. For example, the capability identification matrix is an appropriate tool to design the catalog and to identify, define descriptive elements and describe appropriate criteria as well. Some of them are: *Nested Cylinders and spheres, Cubes, 3D Scatterplots, Net layer models/charts, Tree layer models, Parallel coordinates/matrices, Cluster maps, Portfolios* (Fig. 12.24).

They have to be selected in relation to their respective addressees and degree of detail (Table 12.14).

12.4.3.2 Working Step 2: Capability Content Engineering

Within this step identified capabilities are described in further detail. The catalog's structures are depicted with the help of models that support a clear and consistent conception of the catalog. Prior to any adjustment, a review of previous work is required. Hence, the refinement or renewing of descriptive elements can be initiated.

Therefore, the following questions should be answered for each specific capability:

- *Which are the related, underlying and/or linked roles, processes, departments and capabilities?*
- *Which information is needed as input?*
- *Which resources are needed as input?*
- *Which information and resources are needed?*
- *How can these sources be provided?*

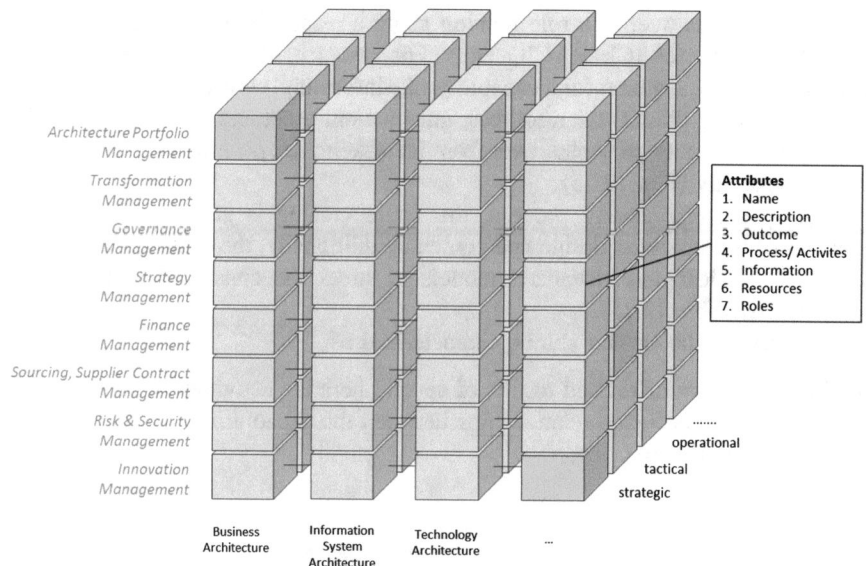

Attributes
1. Name
2. Description
3. Outcome
4. Process/ Activites
5. Information
6. Resources
7. Roles

operational

tactical

strategic

Business
Architecture

Information
System
Architecture

Technology
Architecture

...

Architecture Portfolio
Management

Transformation
Management

Governance
Management

Strategy
Management

Finance
Management

Sourcing, Supplier Contract
Management

Risk & Security
Management

Innovation
Management

Fig. 12.24 Example for a content layers cube from the EACN project (cp. [26, 30])

Table 12.14 BB3.WS1—Summary

Input	Throughput	Output
Capability catalog v1.0	Content layer definition	Capability catalog v2.0 incl. content layer concept

- *What does the capability action in practice look like?*
- *Are there common accepted activities, business processes and responsible roles regarding each capability?*
- *Are there under-/over performing and/or missing capabilities (gap analysis) based on performance targets derived from the strategy?*
- *Which relevant metrics and/or key performance indicators (derived from strategic objectives) can be identified?*

  - *How are they scored (e.g. in terms of properties of the EA to which the capability is linked)?*

- *How can we link capabilities to their motivation (strategic goals)?*
- *How can we link capabilities to their implementation (e.g. descriptive elements as represented by EA models)?*
- *Do we have to involve any experts and stakeholder? Which ones?*
- *Is there any estimation about the needed time and resources?*

Domain experts and manager must be involved in order to give specific inputs. Depending on its expertise it is desirable and certainly possible that these stakeholders pass over to a role relation using a responsibility assignment matrix (RACI, PACSI, RASCI, RACIQ etc.). For instance, if a stakeholder provides activities that are required for a capability like informing, consulting, accountability or responsibility (RACI) for resources, information, processes or company specific descriptive elements, they pass over from its stakeholder position to a role element of the capability (Fig. 12.25).

During the engineering process, the entire catalog is subject of substantial changes of the structure, design and content. Additionally, the catalog's structure is depicted by clear and visualized models to stress the consistent understanding (Fig. 12.26; Table 12.15).

The state of the catalog can be characterized as:

- Representation of revised results of several iterative activities
- Detailed and accepted relationships between the capabilities
- High level of detail.

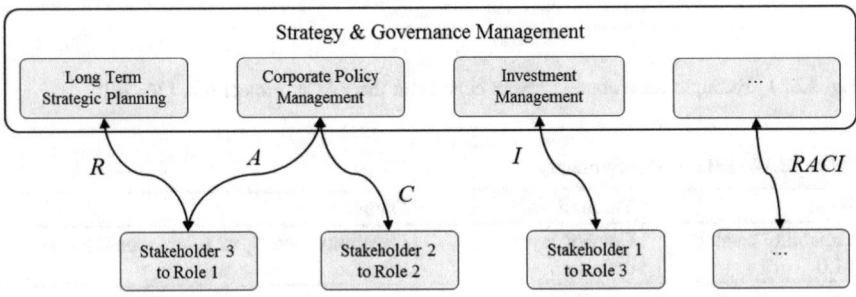

Fig. 12.25 Example for a capability stakeholder-to-role map using RACI assignment

Fig. 12.26 Documentation of content layer 1 and layer 2 within the capability identification matrix

Table 12.15 BB3.WS2—Summary

Input	Throughput	Output
Capability catalog v2.0	Detailing the descriptive elements and defined content layer	Capability catalog v3.0

12.4.3.3   Working Step 3: Development of Stakeholder Views

When describing capabilities in detail, it is necessary to ensure that every capability is formulated in a general manner, i.e., there should not be any connections to objects such as particular applications or markets. In general, views might be applied to present specific sets of capabilities to different kinds of stakeholder groups. In particular, one of the following sample views could be created:

- required maturity level vs. current maturity, level of a capability used for strategy implementation,
- costs of bringing a theoretical capability (some of the descriptive elements are described but nonexistent in the enterprise) into real,
- dependencies between capabilities,
- capabilities required for a particular strategy implementation
- financial aspects,
- just a capability overview map.

For presentation purposes, different tools and technical measures may be used. Different kinds of evaluation criteria are developed in this working step. When describing capabilities in detail, it is necessary to ensure that every capability content layer and its defined elements are formulated and may be linked to other logical elements of the EA.

Procedural Example—The connection between goals, strategies, initiative and corresponding capabilities for realization could be captured in a view (Fig. 12.27).

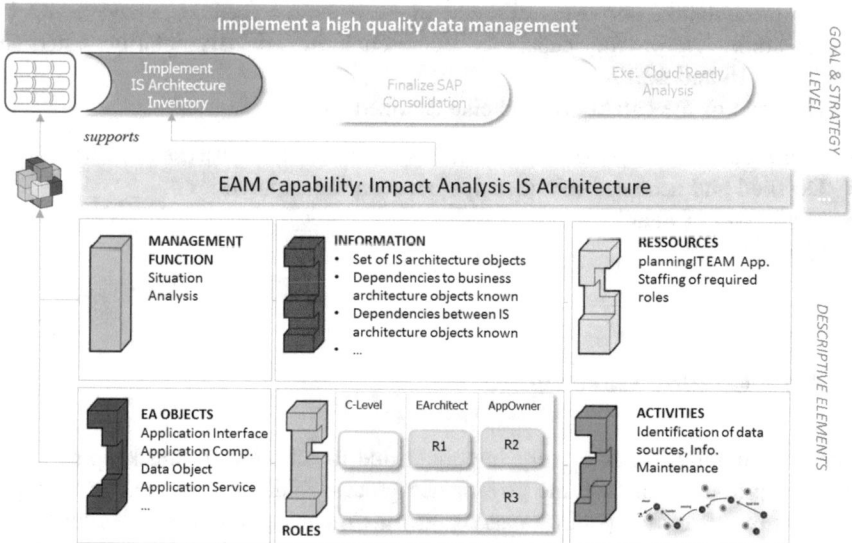

Fig. 12.27 Example impact analysis application architecture capability

Table 12.16 Views for stakeholder specific capability representation

Strategic views	Information views
Performance charting	Radar charts
Strategy maps	Cone tree maps
Feedback diagrams	Hyperbolic tree
Magic quadrants	Cycle diagram
Spray diagrams	
Affinity diagrams	
Mintzberg's organigraph	
Strategic game boards	
Concept views	Compound views
Mind map	Graphic facilitation
Square of oppositions	Rich picture
Concentric cycles	Knowledge map
Synergy map	Informural
Force field diagram	Learning map
Argumentation map	
Perspective diagram	

Table 12.17 BB3.WS3—Summary

Input	Throughput	Output
Capability catalog v3.0	View models simple measurement methods and tools	Capability catalog v4.0

For presentation purposes, different tools, technical methods views could be used. Some views for capability representation are exemplarily listed in Tables 12.16 and 12.17.

The state of the catalog can be characterized as:

- Representation of revised results after several iterative process activities
- Detailed and accepted relationships between the capabilities
- High level of detail
- Completed.

12.4.4 Catalog Governance

This last building block is very important due to introducing and keeping capabilities up-to-date. In fact, the governance process addresses the quality management of the created capability catalog. It thus includes activities referring to the assessment, deployment, and maintenance of a catalog illustrated in Fig. 12.28.

Even though there are a lot of approaches dealing with quality criteria and valuation methods in the context of, for example, enterprise architectures [19].

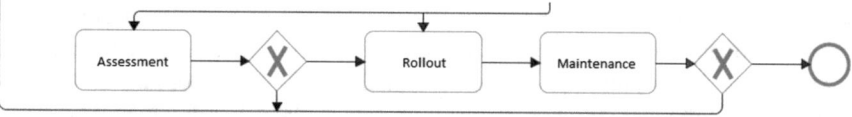

Fig. 12.28 BB4—overview working steps

There is still little progress in the application area of evaluating capabilities, in which approaches most often build on ordinary methods for quality control or are impractical for the designated purpose. This might have originated from an omitted preparation phase, which is normally used to describe the quality criteria a catalog has to satisfy.

12.4.4.1   Working Step 1: Assessment

In order to both counteract deficient quality and promote the functionality of a catalog, the optional step "assessment" can be used. The focus of the assessment concept can be the development process (the way the catalog is constructed), the designed result (the catalog itself), or both. Accordingly, the quality level and quality criteria have to be elaborated during this stage. Appropriate criteria can normally be derived from the goals predefined in BB1 or from the deeper content layers defined and formulated in BB3. In addition to conducting an overall review of general quality standards such as completeness, accuracy, flexibility, linkage, simplicity, intelligibility, and usability, it is recommended to apply comprehensive assessment approaches, e.g., capability maturity models or capability assessment matrices. From the process perspective after such an assessment phase, it is possible to revisit the second and third BB in order to integrate assessment results in a new iteration. Moreover, if the assessment results are absolutely not satisfactory or primary goals are not achieved, the first BB has to be visited again in order to analyze critical points and re-define scope, definitions, stakeholder groups or development strategy. There are three kinds of proposed subject:

- Development process—the way the catalog was constructed
- Engineering result(s)—the content of catalog itself
- Both

   The indicative questions are:

- *What is the objective of the assessment?*
- *Should the catalog be verified (theory-oriented) or validated (practice-oriented)?*
- *Are there any existing quality criteria (e.g. from BB1 or BB3)?*
   - *No: What criteria should we use?*

   There are several opportunities for assessment. 1st: If a **maturity model** is used. Maturity models are specific management instruments, which define various degrees of maturities in order to evaluate to what extent a particular competency fulfils the qualitative requirements that are defined for a set of competency objects [25] and the development processes in organisations [4]. According to Wißotzki et al. [26, 30] the utilization of maturity models in the capability context comprehends three different variants: descriptive, prescriptive or comparative. The descriptive maturity models could be applied to asses the current state (as-is) of a capability or a capability group. Prescriptive models do not only assess the as-is situation, but also recommend guidelines, best practices and roadmaps in order to reach higher degrees of capability maturity. A comparative maturity model can be applied for benchmarking across different capabilities. Following questions can guide the maturity assessment step:

- *For each phase/maturity state*:

    - *Which kind of maturity approach do we need?*
    - *What criteria should they assess?*
    - *Which kind of maturity states/levels do we need?*
    - *How can we close gaps between current and desired states/levels?*

Procedural example—The current data quality of capability can be illustrated by a spider chart (Fig. 12.29). The chart shows significant gaps between the desired level (100 %) and the current level of the required descriptive elements.

2nd: **Portfolios** are a good possibility to compare different assessment criteria like the mapping of investments to existing capabilities (e.g. internal and external distribution; human resource, IT and procure management) illustrated in Fig. 12.30.

Using highlights and/or priorities for a third criterion (e.g. their importance regarding the global enterprise goals) the same portfolio can offer gaps and misleading investments. Consequently, future measurements can be derived due to increase investment in capabilities of low strategic importance and the other way around.

Fig. 12.29 Descriptive elements—maturity visualization

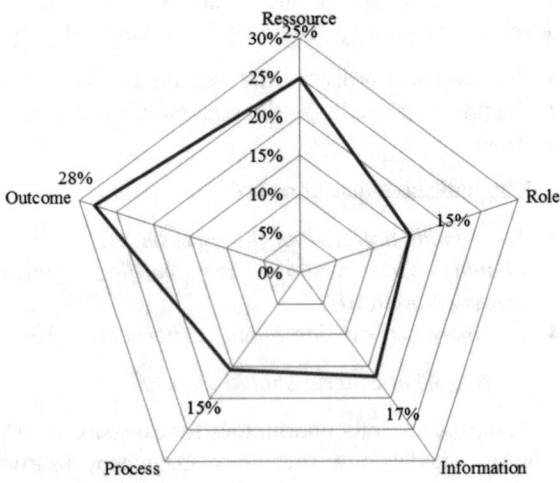

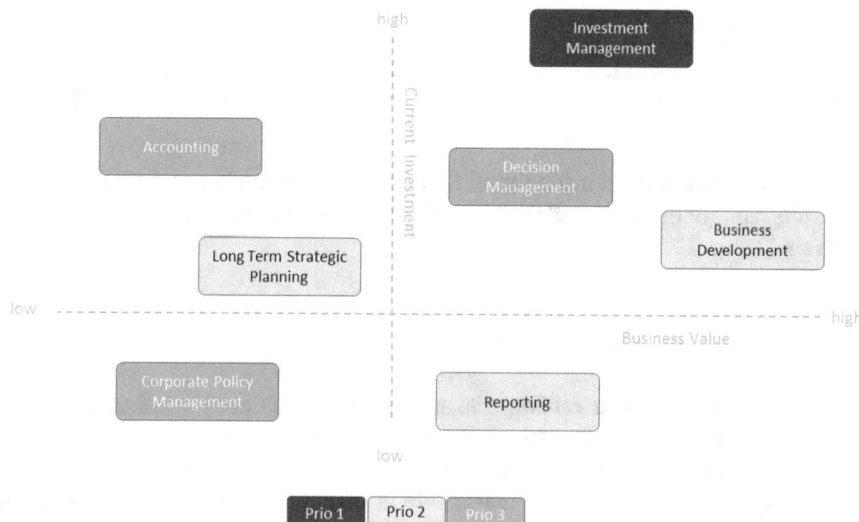

Fig. 12.30 Example for a investment/capability portfolio

Table 12.18 Evaluation matrix

Application	1			
Solution Bulding Block				4
Business Object		2		
Service (abgrenzen zu Appl.)			3	
Application Deployment	1			

However, the quality level of the assessment depends on appropriate chosen assessment criteria. We distinguish between general and specific criteria. General criteria are generally applicable for all capability type. Some examples: completeness, accuracy, flexibility, linkage, simplicity, reasonability, intelligibility, usability, availability or system support.

Specific criteria are capability specific quality indicators and have to be individually defined for each capability. In terms of our EAM Capability "Impact Analysis Application Architecture" driving questions for the quality of its descriptive element "information" could be: Inventory of AA architecture objects available? Are dependencies between AA architecture objects known? Are dependencies to business architecture objects known? Are dependencies to technology architecture objects known? These questions could be answered by a metric like in Table 12.18.

Table 12.19 BB4.WS1—Summary

Input	Throughput	Output
Capability catalog v4.0	Evaluation, indicators, measurements	Capability catalog v4.1

This evaluation can be made for all descriptive element of a capability. The sum of results can, for example, reflect the level of maturity or taken as base for variance analysis (Table 12.19).

12.4.4.2 Working Step 2: Rollout

The way of integrating a catalog into an enterprise has a vital influence on the success of this catalog. To this end, the "catalog rollout" step addresses the implementation/rollout of a catalog in the organization. As specified earlier, creating a capability catalog is only reasonable in case the management approves and supports the process. Accordingly, both upper and middle management need to be convinced. That being said, the success of integrating a capability catalog depends on two major elements: quality and stakeholder satisfaction.

The completed capability catalog thus needs to be formally presented to the steering committee and contracting authority, respectively. This should be delivered either in the form of an intermediate presentation or as part of the project completion. It thus needs to be ensured that the needs of the stakeholders are satisfied. To achieve this, accurate planning and preparation is required. The project team needs to be able to enhance the results of the capability catalog creation process, i.e., converting the final catalog version, descriptions, and illustrations into an appropriate form of presentation. Relevant stakeholders might, for example, obtain a copy of the document in order to prepare themselves for approval.

All in all, the catalog rollout needs to pursue the goal of achieving an acceptance of the results and creating an activity plan in terms of additional elaborations or unresolved issues. Even though an initial evaluation of the achieved state should have been conducted in the preceding building blocks, it is unlikely that a single iteration is sufficient. The second goal is to receive user feedback provided by individuals or working groups in order to improve the catalog utilization. In this regard, it is recommended to perform internal surveys or workshops after a certain period of time. The way of integrating a catalog into an enterprise has a vital influence on the success of it. Two major elements influence the integration significantly: (1) The capability catalog has a high-quality level, (2) Stakeholders (e.g. board level, business developers, line managers) are satisfied with both the approaches and achieved results.

Table 12.20 BB4.WS2—Summary

Input	Throughput	Output
Capability catalog v4.1	Integration	Capability catalog v4.2

Three subsequent aspects need to be considered in the context of catalog deployment: Obtain feedback from users and steering committee, obtain decisions about the maintenance of the catalog and the allocation of resources, integrate the catalog into existing standards (Table 12.20).

12.4.4.3   Working Step 3: Maintenance

Feedback from the previous working step or especially from catalog utilization can result in a change in the structure and/or in the function of catalog elements. Besides, changes in the enterprise (e.g. governance, new orientation, management) and its branch can create the need for improvements in the catalog. For these reasons, and given that an enterprise may have to meet new challenges over time and capabilities need to be modified accordingly, there is an ongoing "maintenance" process in addition to the aforementioned evaluation methods applied to create a high-quality capability catalog. Consequently, an improvement of both quality and usage period of the catalog is addressed within the last step of this building block. Modifications in the catalog structure as well as slight changes may occur in this step.

Due to changing challenges for enterprises over time, capabilities need to be modified accordingly. With the collaboration of the optional, aforementioned evaluation methods, an ongoing maintenance results in a high-quality capability catalog.

Which measures and modifications improve the catalog quality?

The including advantages are:

- Structure and comprehensibility
- Precise descriptions
- Simplified modifications and reorganizations of the created catalog
- Contributes to the organizational learning and securing of organizational knowledge
- An improvement of both usage period and quality of the catalog is addressed.

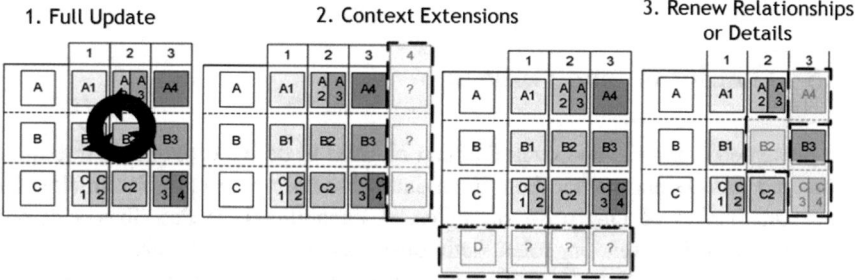

Fig. 12.31 Capability catalog maintenance patterns according to Lahrmann and Marx [17]

Table 12.21 BB4.WS3—Summary

Input	Throughput	Output
Capability catalog v4.2	Maintenance activities	Capability catalog v5.0

From Lahrmann and Marx [17], we adopted three of four extension patterns for the purpose of catalog maintenance (Fig. 12.31).

A general update of capability catalog elements by adding new descriptive elements or updating the evaluation mechanism (e.g., maturity assessment procedure) may be examples of the first pattern. It is also possible to add new context objects or reorder their configurations, e.g., by changing attributes that might influence the identification process (*BB2*) or at least reconfigure the relationships between different capabilities. Although these extension patterns challenge the meta-structure of the capability catalog to some extent, they would not require passing the first building block and beginning the development process again by redefining the scope, as this would go beyond the scope of maintenance. All stakeholders (initiators, developers and users) have to maintain the existing catalog and influence its improvement (all-do-some approach), if there are any needs of change.

Is an update or upgrade necessary?

The differentiation between upgrade and update is in the point of interest.

1. Update: adding new descriptive elements or content objects; updating the evaluation methods; any kind of modifications (BB2–BB4)
2. Upgrade: total renewing of the catalog starting with BB1

Procedural example—When enterprise roles changed or processes are basically modified (f.i. by automatization), the existing catalog has to be updated by new iteration in terms of identifying additional or obsolete capabilities starting from *BB2* to *BB4*. An organizational restructuring or market change can lead to a capability catalog upgrade starting from re-scoping in BB1–BB4.

The existing capability catalog has to be: Modified regarding the steps of *BB2* till *BB4* and comply with the described state of the art **OR** totally renewed from *BB1*. Then the existing capability catalog is rejected and a new one, corresponding to the single BB's and working steps, is designed (Table 12.21).

12.5 Conclusion and Outlook

We presented a generic approach that can be used to derive capabilities through a structured process and gather them in an enterprise-specific catalog for an effective operationalization of enterprise strategies. A capability here describes a certain combination of information, roles, activities/procedures, and resources to support issues like strategy management, enterprise architecture management and IT management considering EA objects and management tasks like planning, transformation and controlling.

Following a four-building-block approach, we described a straightforward and flexible process for capability catalog developers and designers, which allows the integration of descriptive elements for different capability types. The capability management guide version 3 is based on the approach of Wißotzki [28] and includes changes of two expert evaluations. It should form a tool that facilitates the development of scientifically well-founded capability catalogs aligned with the design science research guidelines [16] and process [18]. In particular, our approach provides a building block covering the continuous assessment and maintenance in order to enhance capability and catalog quality.

Additional detailed content of the building blocks and corresponding steps are still in development and have only been mentioned to some extent in this chapter. Our future research will elaborate on this topic and demonstrate more practical use cases of capability catalog engineering projects. In fact, our aim is to focus more on use cases and/or possible applications in order to indicate the tradeoffs of our approach and to evaluate and potentially extend the process.

References

1. Abowd, G.D., Dey, A.K., Brown, P.J., Davies, N., Smith, M., Steggles, P.: Towards a better understanding of context and context-awareness. In: Proceedings of the 1st International Symposium on Handheld and Ubiquitous Computing, Karlsruhe, Germany, 27–29 Sept 1999
2. Ahlemann, F., Stettiner, E., Messerschmidt, M., Legner, C. (eds.): Strategic Enterprise Architecture Management: Challenges, Best Practices, and Future Developments. Springer, Berlin (2012)
3. Andreu, R., Ciborra, C.: Organisational learning and core capabilities development: the role of it. J. Strateg. Inf. Syst. 5, 111–127 (1996)
4. Back, A.: Reifegradmodelle im Management von Enterprise 2.0. BITKOM (2010)
5. Bharadwaj, A.: A resource-based perspective on information technology capability and firm performance: an empirical investigation. MIS Q. 24(1), 169–196 (2000)
6. Buckl, S., Dierl, T., Matthes, F., Schweda, C.M.: Building blocks for enterprise architecture management solutions. In: Harmsen F., Proper E., Schalkwijk F., Barjis J., Overbeek S. (eds) Practice-driven research on enterprise transformation, lecture notes in business information processing, vol. 69, pp. 17–46. Springer, Heidelberg (2010)
7. Capgemini Application Landscape Report 2011: Radar exosystems specialists: Whitepaper. The impact of data silos on IT planning (2012)
8. CEB CIO Leadership Council (ed): Get business capabilities right. CEB CIO Leadership Council (2015)
9. Day, G.S.: The capabilities of market-driven organizations. J. Mark. 58(4), 37–52 (1994)
10. Dosi, G., Nelson, R.R., Winter, S.G.: The nature and dynamics of organizational capabilities. Introduction: The Nature and Dynamics of Organizational Capabilities, pp. 1–22. Oxford University Press, Oxford (2000)
11. Espana, S., Grabis, J., Henkel, M., Koc, H., Sandkuhl, K., Stirna, J., Zdravkovic, J.: Strategies for capability modelling: analysis based on initial experiences. In: Persson A., Stirna J. (eds) CAiSE 2015 Workshops, LNBIP 215. Springer, Berlin (2015)
12. Fischer, R.: Organisation der Unternehmensarchitektur: Entwicklung der aufbau-und ablauforganisatorischen Strukturen unter besonderer Berücksichtigung des Gestaltungsziels Konsistenzerhaltung. Univ, Diss.–Sankt Gallen, 1st ed. Verlag Dr. Kovac (2008)

13. Freitag, A., Matthes, F., Schulz, C., Nowobilska, A.: A method for business capability dependency analysis. In: International Conference on IT-enabled Innovation in Enterprise (ICITIE2011), Sofia (2011)
14. Gartner Inc.: Technology issues for financial executives: 2011 annual report. Gartner and Financial Executives International (FEI), Morristown, NJ (2011)
15. Greski, L.: Business capability modeling: theory & practice. Architecture & Governance Magazine, 22 Dec 2009
16. Hevner, A.R., March, S.T., Park, J., Ram, S.: Design science in information systems research. MIS Q. **28**(1), 75–106 (2004)
17. Lahrmann, G., Marx,F.: Systematization of maturity model extensions. In: Winter R., Zhao J. L., Aier S. (eds) Global Perspectives on Design Science Research. Lecture Notes in Computer Science, vol. 6105, pp. 522–525. Springer, Berlin (2010)
18. Peffers, K., Tuunanen, T., Rothenberger, M.A., Chatterjee, S.: A design science research methodology for information system research. J. Manage. Syst. (JMIS) **24**(3), 45–77 (2008)
19. Sandkuhl, K., et al.: Enterprise Modeling. Tackling Business Challenges with the 4EM Method, 309, Springer (2014)
20. Simon, D., Fischbach, K., Schoder, D.: Enterprise architecture management and its role in corporate strategic management. IseB **12**(1), 5–42 (2014)
21. Stutz, M.: Kennzahlen für Unternehmensarchitekturen – Entwicklung einer Methode zum Aufbau eines Kennzahlensystems für die wertorientierte Steuerung der Veränderungen von Unternehmensarchitekturen. Univ, Diss.–Sankt Gallen, 1st ed. Verlag Dr. Kovac, Hamburg, Germany (2009)
22. Ulrich, D., Smallwood, N.: Capitalizing on capabilities. Harvard Business Review on Point, 12 June 2004
23. Ulrich, W., Rosen, M.: The business capability map: the "Rosetta Stone" of business/IT alignment. Cut. Consort. Enterp. Archit. **14**(2) (2011)
24. Weldon, L., Burton, B.: Use Business Capability Modeling to Illustrate Strategic Business Priorities. Gartner Inc. (2011)
25. Wendler, R.: Reifegradmodelle für das IT-Projektmanagement. Techn. Univ. Fak. Wirtschaftswiss, Dresden (2009)
26. Wißotzki, M., Koç, H., Weichert, T., Sandkuhl, K.: Development of an enterprise architecture management capability catalog. In: Kobyliński A., Sobczak A. (eds) Perspectives in Business Informatics Research. Lecture Notes in Business Information Processing, vol. 158, pp. 112–126. Springer, Berlin, Heidelberg (2013)
27. Wißotzki, M., Sandkuhl, K.: Elements and characteristics of enterprise architecture capabilities. In: Perspectives in Business Informatics Research, pp. 82–96. Springer International Publishing (2015)
28. Wißotzki, M.: The capability management process—finding your way into capability engineering. In: Simon D., Schmidt C. (eds) Business Architecture Management— Architecting the Business for Consistency and Alignment. To be published by Springer in the series "Management Professionals" (2015)
29. Wißotzki, M., Koç, H.: Evaluation concept of the enterprise architecture management capability navigator. In: 16th International Conference on Enterprise Information Systems (ICEIS 2014), Lisbon, Portugal (2014). ISBN: 978-989-758-027-7
30. Wißotzki, M., Koç, H., Weichert, T.: A project driven approach for enhanced maturity model development for EAM capability evaluation. In: Bagheri E. (ed) 17th IEEE International Enterprise Distributed Object Computing—Conference, pp. 296–305. Vancouver and Canada (2013). ISBN: 978-0-7695-5085-5

Author Index

© Springer International Publishing Switzerland 2016   265
E. El-Sheikh et al. (eds.), *Emerging Trends in the Evolution of Service-Oriented
and Enterprise Architectures*, Intelligent Systems Reference Library 111,
DOI 10.1007/978-3-319-40564-3

nited States
sters